Recouping Separation Pay from U.S. Service Members and Veterans Who Later Receive Veterans Affairs Disability Compensation

STEPHANIE RENNANE, BETH J. ASCH, MICHAEL G. MATTOCK,
HANNAH ACHESON-FIELD

Prepared for Office of the Under Secretary of Defense
for Personnel and Readiness
Approved for public release; distribution unlimited

NATIONAL DEFENSE RESEARCH INSTITUTE

For more information on this publication, visit **www.rand.org/t/RRA851-1**.

About RAND

The RAND Corporation is a research organization that develops solutions to public policy challenges to help make communities throughout the world safer and more secure, healthier and more prosperous. RAND is nonprofit, nonpartisan, and committed to the public interest. To learn more about RAND, visit www.rand.org.

Research Integrity

Our mission to help improve policy and decisionmaking through research and analysis is enabled through our core values of quality and objectivity and our unwavering commitment to the highest level of integrity and ethical behavior. To help ensure our research and analysis are rigorous, objective, and nonpartisan, we subject our research publications to a robust and exacting quality-assurance process; avoid both the appearance and reality of financial and other conflicts of interest through staff training, project screening, and a policy of mandatory disclosure; and pursue transparency in our research engagements through our commitment to the open publication of our research findings and recommendations, disclosure of the source of funding of published research, and policies to ensure intellectual independence. For more information, visit www.rand.org/about/principles.

RAND's publications do not necessarily reflect the opinions of its research clients and sponsors.

Published by the RAND Corporation, Santa Monica, Calif.
© 2022 RAND Corporation
RAND® is a registered trademark.

Library of Congress Cataloging-in-Publication Data is available for this publication.

ISBN: 978-1-9774-0721-4

Cover: Retired Army 1st Sgt. William Staude, of Elliott, Pa., salutes soldiers from the 316th Expeditionary Sustainment Command, stationed in Coraopolis, Pa., as they march past him during the Veterans Day parade in downtown Pittsburgh, Nov. 11. Photo by Sgt. 1st Class Michel Sauret/U.S. Army.

About This Report

When service members separate from the military, they may receive one of many types of separation benefits. Some of these service members later become eligible for other types of compensation, such as U.S. Department of Veterans Affairs (VA) Disability Compensation (VADC). To avoid paying more than one award of compensation to any person based on that individual's own service—what is sometimes called *double-dipping*—VA is prohibited from paying compensation to a veteran who also received certain separation benefits from the U.S. Department of Defense (DoD) until the full separation amount has been withheld, or *recouped*.

Because of concern about how recoupment affects veterans, Section 654 of the National Defense Authorization Act of 2020 required the Secretary of Defense, in consultation with the Secretary of Veterans Affairs, to conduct a study to determine

1. how many members and veterans who received separation pay, special separation benefits, and voluntary separation incentive payments are affected by recoupment
2. the aggregated amount of additional money members and veterans would receive in the absence of recoupment.

To assist with this study, the Office of the Under Secretary of Defense for Personnel and Readiness requested analytic support from the RAND Corporation's National Defense Research Institute. This report summarizes the findings of the study.

RAND National Security Research Division

The research reported here was completed in June 2021 and underwent security review with the sponsor and the Defense Office of Prepublication and Security Review before public release.

This research was sponsored by the Office of the Under Secretary of Defense for Personnel and Readiness and conducted within the Forces and

Resources Policy Center of the RAND National Security Research Division (NSRD), which operates the National Defense Research Institute (NDRI), a federally funded research and development center sponsored by the Office of the Secretary of Defense, the Joint Staff, the Unified Combatant Commands, the Navy, the Marine Corps, the defense agencies, and the defense intelligence enterprise.

For more information on the RAND Forces and Resources Policy Center, see www.rand.org/nsrd/frp or contact the director (contact information is provided on the webpage).

Acknowledgments

We thank many individuals who contributed to and supported this study. First, we thank our sponsor, Lernes J. Hebert, Deputy Assistant Secretary of Defense for Military Personnel Policy. We would also like to thank our project monitors, CDR David Clark and LCDR Ryan Roeling, in Military Compensation Policy in the Office of the Under Secretary of Defense for Personnel and Readiness (OUSD[P&R]), as well as Jeri Busch, Director of Military Compensation Policy in OUSD(P&R), for sharing insight and guidance on this report. We are grateful to many individuals at the Defense Finance and Accounting Service and the U.S. Department of Veterans Affairs (VA) who provided invaluable assistance with data access, including Brian Mathewuse (VA). We thank our RAND Corporation colleagues John Winkler, Molly McIntosh, and Craig Bond for their help and support with this report. We thank Carrie Farmer and Matt Goldberg for serving as technical peer reviewers.

Summary

Service members and veterans receive many types of compensation for various purposes over their careers. When service members separate from the military, they may receive one of many types of separation benefits. These benefits are typically based on the amount of time served in the military. Some service members who receive separation pay later become eligible for U.S. Department of Veterans Affairs (VA) Disability Compensation (VADC). To avoid paying more than one award of compensation to any person based on that individual's own service (38 U.S.C. 5304(a))—what is sometimes known as *double-dipping*—VA is prohibited from paying compensation to a veteran who also received certain separation benefits from the U.S. Department of Defense (DoD) until the full separation pay amount has been withheld. This process is referred to as *recoupment*. Veterans may be subject to recoupment if they qualify for VA compensation at any point after receiving separation benefits.

Recoupment of separation pay is complicated and can cause confusion and frustration among veterans (Clark, 2017; Tilghman, 2015). On the one hand, from the standpoint of 38 U.S.C. 5304(a), veterans are not losing money, because they were paid previously in the form of separation pay and a deduction is being made to avoid duplication of benefits. On the other hand, veterans might not know that their VA benefits will be reduced if they received separation pay in the past, and recoupment can come as a surprise if the veteran qualifies for VA compensation many years after receiving separation pay. Veterans might perceive separation pay as being completely disconnected from VADC. This might be particularly true in the case of veterans who took the voluntary separation incentive (VSI) or special separation benefits (SSB). These payments were intended to encourage members to separate voluntarily to help the services draw down end strength in the 1990s. Furthermore, because separation pay can be thousands of dollars, it can take years to fully recoup. In the meantime, veterans with disabilities do not receive disability compensation from VA. These issues have resulted in many inquiries and questions to Congress about recoupment of separation pay.

Because of concern about how recoupment affects veterans, Section 654 of the National Defense Authorization Act for Fiscal Year 2020 required the Secretary of Defense, in consultation with the Secretary of Veterans Affairs, to conduct a study to determine

1. how many members and veterans who received separation pay, SSB, and VSI payments are affected by recoupment
2. the aggregated amount of additional money members and veterans would receive in the absence of recoupment (Pub. L. 116-92, 2019).

To assist with this study, the Office of the Under Secretary of Defense for Personnel and Readiness requested analytic support from the RAND Corporation's National Defense Research Institute. This report summarizes the findings of the study.

Research Approach

We analyzed data provided by VA and the Defense Finance and Accounting Service (DFAS). VA data provided information on recoupment of separation pay (voluntary separation pay [VSP] or involuntary separation pay [ISP]) and SSB for those who experienced recoupment because of receipt of VA benefits in fiscal year (FY) 2013 or later. The DFAS data provide information on recoupment from VSI payments due to receipt of VA benefits in FY 1992 or later. DFAS also provided data on recoupment of separation pay due to the receipt of military retired pay. Consequently, we supplemented our analysis of recoupment due to receipt of VA benefits with analysis of recoupment due to receipt of DoD retired pay.

We conducted a series of tabulations to respond to the two questions posed by the congressional request. We quantified the total number of members and veterans who are subject to recoupment of separation pay from VA benefits and from military retired pay, subject to the year limitations in our data. We then estimated the amount that would have been received in the absence of recoupment in two ways. First, we computed the average amount of recoupment each member has experienced. Second, we estimated the aggregate total amount of recoupment when summed across all service members.

Main Findings

Table S.1 presents a summary of our results. We estimate that just over 72,000 veterans experienced recoupment of VSP or ISP between 2013 and 2020.[1] Another 2,600 experienced recoupment of SSB between 2013 and 2020, and 4,700 have experienced recoupment of VSI since the program began in 1992. Given the available data, we cannot determine how many veterans began and completed recoupment of VSP, ISP, or SSB prior to 2013. On average, veterans have had $19,700 or $25,700 withheld because of recoupment of ISP/VSP or SSB, respectively, but have had $53,000 withheld because of recoupment of VSI. In aggregate, over the eight-year period of 2013 to 2020, a total of $1.4 billion in VADC payments were withheld because of the recoupment of VSP or ISP, a far larger figure than the $68 million withheld because of the recoupment of SSB. DFAS has withheld a minimum of $177 million of VSI in aggregate because of the receipt of VADC.

In considering these results, it is important to note several data limitations. In particular, the data did not permit us to compute the total number

TABLE S.1

Summary of Results for the Two Congressional Questions

Separation Pay Type	Question 1	Question 2		
	Count of Veterans	Average Recouped per Person	Total Recouped	Estimated Total Recouped in 2021
VSP or ISP (2013–2020)	72,206	$19,700	$1.4 billion	$67 million
SSB (2013–2020)	2,651	$25,700	$68 million	$730,000
VSI (1992–2020)	4,701	$53,000	$177 million	$9.6 million

SOURCES: Data from 1992 to 2020 provided to the authors by DFAS; data from 2013–2020 provided to the authors by VA.
NOTE: VSI computations for Question 2 are based on cases in which recoupment has been completed. Estimates of the total amount to be recouped in 2021 are based on cases in which we observed recoupment at the time that the data were collected. The total amount recouped for VSI is based on cases in which VSI payments are complete.

---

[1]   For ease of exposition, we use the term *veterans* even though the congressional language uses *members and veterans*.

of veterans who experienced and completed VA recoupment of either SSB or separation pay before 2013 because VA data systems underwent a major change at that time. Furthermore, we were unable to compute either the share of service members who received SSB or VSP/ISP who ever experienced VA recoupment or the share of veterans receiving VADC who ever experienced VA recoupment. Finally, we received separate information for service members who experienced and completed recoupment of VSI and for those still experiencing recoupment of VADC from VSI. Because of differences in the structure of the data sources, we were unable to determine the total amount recouped before December 2020 for cases that are still experiencing recoupment.

Contents

About This Report iii
Summary v
Figures and Tables xi

CHAPTER ONE
Introduction 1

CHAPTER TWO
Background on Recoupment of Separation Pay 5
    VA Disability Compensation 5
    Types of Separation Pay Subject to Recoupment 6
    Recoupment 11
    Trends in Separations with Involuntary Separation Pay, Voluntary
        Separation Incentive, and Special Separation Benefits 15

CHAPTER THREE
Data and Research Approach 17
    Data on Recoupment from VA Disability Compensation 17
    Data on Recoupment from Military Retired Pay 21
    Research Approach 22

CHAPTER FOUR
Results 25
    Question 1: How Many Members and Veterans Are Affected by
        Recoupment of Separation Pay? 25
    Question 2: What Is the Aggregated Amount of Additional Money
        Members and Veterans Would Receive in the Absence of
        Recoupment? 29

CHAPTER FIVE
Findings and Discussion 45
    Findings 45
    Discussion 46
    Wrap-Up 51

Appendix 53
Abbreviations 59
References 61

Figures and Tables

Figures

1.1. Conceptual Overview of Recoupment of Separation Pay Because of VADC and Because of Receipt of Military Retired Pay.. 3

2.1. Trend over Time in Number of Separating Personnel Receiving ISP... 15

3.1. Observed and Estimated Recoupment of VSI for Completed and Ongoing Payments 19

4.1. Count of Veterans with Any Recoupment of Separation Pay (VSP or ISP) or SSB Due to Receipt of VADC Between 2013 and 2020 26

4.2. Count of Veterans with Recoupment of VSI Due to Receipt of VADC 27

4.3. Count and Cumulative Share of Veterans with Outstanding VSI Recoupment, Based on Completion Year of VSI and Recoupment....................................... 28

4.4. Average Per-Person After-Tax Value of Separation Pay (VSP or ISP) and SSB Among Awards Subject to Recoupment, 2013–2020... 31

4.5. Aggregate Total Amount of VSP or ISP Recouped, 2013–2020... 33

4.6. Aggregate Total Amount of SSB Recouped, 2013–2020 34

4.7. Annual Amount Expected to be Recouped from VSI by Year, and Cumulative Share of Outstanding Amounts........... 37

4.8. Count of Veterans with Any Recoupment of Separation Pay, SSB, or VSI Due to Receipt of Military Retired Pay Between 2013 and 2020 39

4.9. Average Per-Person Amount Recouped to Date from Separation Pay, SSB, and VSI Among Awards with Ongoing Recoupment from Military Retired Pay 41

A.1. Number of Veterans with Any Recoupment of Separation Pay (VSP or ISP), SSB, or VSI Due to Receipt of Military Retired Pay over Time, 2013–2020 53

A.2.   Number of Veterans with Full Recoupment from Military
       Retired Pay Observed Between 2013 and 2020, by Type of
       Separation Pay 54
A.3.   Average Per-Person Amount of Separation Pay, SSB,
       and VSI Recouped, Among Awards with Completed
       Recoupment from Military Retired Pay 55
A.4.   Count of Cases of Recoupment of SSB from Military
       Retired Pay, Based on the Number of Years Required to
       Fully Recoup SSB 56
A.5.   Count of Cases of Recoupment of ISP or VSP from Military
       Retired Pay, Based on the Number of Years Required to
       Fully Recoup Separation Pay 57
A.6.   Count of Cases of Recoupment of VSI from Military
       Retired Pay, Based on the Number of Years Required to
       Fully Recoup VSI 58
A.7.   Cumulative Distribution of the Number of Years Required
       to Recoup Separation Pay from Military Retired Pay,
       by Pay Type 58

Tables

S.1.   Summary of Results for the Two Congressional Questions..... vii
2.1.   Types of Separation Pay Subject to Recoupment 8
2.2.   VA Recoupment of Separation Pay 12
2.3.   Number of Separations with SSB or VSI, 1992 to 2001 16
3.1.   Type of Separation Pay and Observed Recoupment in Each
       Data Source 20
3.2.   Information Available for Computing Recoupment for Each
       Type of Separation Pay 21
4.1.   Amounts and Distribution of Separation Pay 30
4.2.   VSI Award and Recoupment Totals Among Cases with
       Completed Recoupment 36
4.3.   Aggregated Total Amount Recouped from Military Retired
       Pay, 2013–2020 43
5.1.   Summary of Results for the Two Congressional Questions...... 45
5.2.   Summary of Results for Recoupment from Military Retired
       Pay, 2013–2020 48

CHAPTER ONE

Introduction

To avoid paying more than one award of compensation to any person based on that individual's military service, Section 5304(a) of Title 38 of the U.S. Code of Federal Regulations prohibits the U.S. Department of Veterans Affairs (VA) from paying compensation to a veteran who also received certain separation benefits from the U.S. Department of Defense (DoD) until the full separation amount has been withheld. For example, a service member might leave the military with separation benefits but later qualify for VA Disability Compensation (VADC). Regulation requires that VA reduces VA compensation by the total amount of the separation pay the individual received so that the individual does not receive dual compensation for a single period of service. VA uses the term *recoupment* to describe the process of deducting separation pay from VA benefits, and different types of separation pay have different rules for recoupment. But, in general, recoupment means that a former service member must have their separation pay fully withheld before they can receive VA compensation.[1]

Recoupment of separation pay can cause confusion and consternation among veterans (Clark, 2017; Tilghman, 2015).[2] From the standpoint of 38 U.S.C. 5304(a), veterans are not losing money, because they were paid previously in the form of separation pay and a deduction is being made

---

[1] As we explain in more detail later, the exception to this statement is the voluntary separation incentive (VSI), in which DoD withholds from annual VSI payments an amount equal to the VADC.

[2] As we describe later in this chapter, our analysis focuses on separation pay programs intended to induce service members to leave the military voluntarily and not on disability severance pay (DSP), another type of separation pay. DSP is given to service members who are deemed unfit for continued service and have a DoD disability rating of less than 30 percent.

to avoid duplication of benefits. However, veterans might be unaware that their VA benefits will be reduced if they received separation pay in the past, and recoupment can come as a surprise if the veteran qualifies for VA compensation many years after receiving separation pay. Veterans might perceive separation pay as being completely disconnected from disability compensation, particularly if they voluntarily took the VSI or special separation benefits (SSB) to help draw down end strength in the 1990s. Furthermore, because separation pay can be thousands of dollars, it can take years before separation pay is fully recouped and before VA compensation can begin without a reduction. In the meantime, veterans with disabilities do not receive full disability compensation from VA. These issues have resulted in many inquiries to Congress about recoupment of separation pay.

Because of concern about how recoupment affects veterans, Section 654 of the National Defense Authorization Act (NDAA) for Fiscal Year 2020 required the Secretary of Defense, in consultation with the Secretary of Veterans Affairs, to conduct a study to determine

1. how many members and veterans who received separation pay, SSB, and VSI payments are affected by recoupment
2. the aggregated amount of additional money members and veterans would receive in the absence of recoupment (Pub. L. 116-92, 2019).

To assist with this study, the Office of the Under Secretary of Defense for Personnel and Readiness requested analytic support from the RAND Corporation's National Defense Research Institute. This report summarizes the findings of the study.

The NDAA specifically mentioned three types of separation benefits. Separation pay is paid to service members who are separated and do not qualify for retirement.[3] SSB and VSI were separation pay programs created in 1992 to facilitate the downsizing of the armed forces following the end of the Cold War by encouraging members to separate voluntarily. As

---

[3] Service members must meet a service requirement to receive separation pay. The requirements associated with separation pay and the details of the SSB and VSI program are described in Chapter Two. Also described are the other separation benefits subject to recoupment.

described in more detail in Chapter Two, SSB was a lump-sum payment while VSI was paid in annual installments where the number of VSI payments equaled twice the number of years of active service. Another important difference between SSB and VSI from the standpoint of recoupment is that VA is responsible for recouping only SSB from a veteran's disability compensation, not VSI. Instead, DoD must reduce the veteran's annual VSI payment to offset the amount of VA compensation the veteran receives in a given year. Thus, both VA and DoD are responsible for recoupment depending on the type of separation benefits.

Receipt of VA benefits is not the only reason why veterans may experience recoupment of separation pay. They may also experience recoupment if they qualify for military retired pay. Title 10 of the U.S. Code requires recoupment of separation pay for members who later qualify for military retirement benefits (DoD, 2021b). In some cases, veterans who received separation pay may qualify for both military retired pay and VA compensation, as shown in Figure 1.1. The green circle shows the population of individuals experiencing recoupment of their separation pay because of VADC, while the blue circle is the population experiencing recoupment because of receipt of military retired pay.

FIGURE 1.1

Conceptual Overview of Recoupment of Separation Pay Because of VADC and Because of Receipt of Military Retired Pay

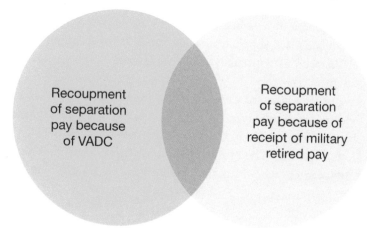

In the intersection of the two circles, veterans may qualify for both military retired pay and VA compensation. In this case, the total amount of recoupment is divided between the two agencies. Typically, the VA recoupment occurs first because service members often qualify for VA disability before they qualify for military retirement. The Defense Finance and Accounting Service (DFAS) reduces the total amount subject to be recouped from military retired pay by the amount recouped by VA so that separation pay is not recouped twice.

The congressional requirement focuses on the population of veterans shown in the green circle. Although recoupment of separation pay from military retired pay (the blue population) was not specifically mentioned in the NDAA reporting requirement, it is relevant to the analysis because of its potential interaction with recoupment of separation pay from VA compensation. Furthermore, insofar as any congressional policy change eliminating recoupment would also affect veterans who experience recoupment from retired pay by DoD, it is of interest to understand recoupment of separation pay from retired pay as contextual background to the analysis.

To estimate the number of members and veterans affected by recoupment and the aggregated additional amount of money they would receive in the absence of recoupment, we analyzed data from VA and DFAS. The VA data provided information on recoupment of separation pay and SSB for those who experienced recoupment because of receipt of VA benefits in fiscal year (FY) 2013 or later. The DFAS data provided information on recoupment from VSI payments in FY 1992 or later. DFAS also provided data on recoupment of separation pay due to the receipt of military retired pay. Consequently, we supplemented our analysis of recoupment due to receipt of VA benefits with analysis of recoupment due to receipt of DoD retired pay.

This report is organized as follows. The next chapter provides a more detailed description of the types of separation pay and the rules for recoupment. Chapter Three describes the data we received from VA and DFAS and our analytic approach. The results of the analysis are presented in Chapter Four. In Chapter Five, we discuss the findings and draw conclusions.

CHAPTER TWO

Background on Recoupment of Separation Pay

This chapter describes the recoupment process. First, we describe the purpose of VADC and eligibility requirements. Second, we list the separation benefits that are subject to recoupment. Third, we put these two together and show the process for VA or DoD recoupment of separation benefits. Finally, we summarize trends in separations from military service with separation benefits.

VA Disability Compensation

Veterans who incurred a service-connected disability are eligible to receive VADC if they separated or were discharged from service under conditions other than dishonorable. VADC is an untaxed, recurring monthly benefit that is intended to offset a reduction in civilian earnings. The amount that eligible veterans receive is based on the number of dependents, relationship of dependents (i.e., child, spouse, or parent), and severity of service-connected disabilities. Veterans who apply can receive a disability rating of between 0 percent and 100 percent, where the rating increases with the severity of the disability. They can also receive multiple ratings if they have more than one qualifying disability; these ratings are then fused into a combined disability rating.[1] Compensation is typically higher if a veteran

---

[1] Individual ratings generally do not combine in a purely additive fashion. For example, a pair of 10-percent disabilities does not necessarily equate to a combined rating of 20 percent.

has more dependents, and compensation increases linearly from a rating of 30 percent to 90 percent and then is further increased at 100 percent (Asch, Hosek, and Mattock, 2014; VA, undated).

VA is responsible for paying VADC to qualifying veterans and for recouping most separation benefits.

Types of Separation Pay Subject to Recoupment

The two main authorities that prohibit the concurrent receipt of multiple forms of compensation and benefits are 38 C.F.R. 3.700 and 38 U.S.C. 5304. The latter statute states that service members cannot receive multiple forms of compensation:

> Not more than one award of pension, compensation, emergency officers', regular, or reserve retirement pay, or initial award of naval pension granted after July 13, 1943, shall be made concurrently to any person based on such person's own service or concurrently to any person based on the service of any other person.

38 U.S.C. 5304 also lays out several exceptions,[2] although it does not specify which agency must recoup payments or how receipt of additional compensation should be recouped. That said, 38 C.F.R. 3.700 specifies how VA should recoup additional compensation; it describes the types of compensation (including separation benefits) that are subject to recoupment and lists the authorities associated with each separation benefit, any applicable dates, and the amount that will be recouped.

Table 2.1 lists the nine different separation benefits that are subject to VA recoupment and provides details about relevant authorities and eligibility criteria. (Note that there are two rows associated with readjustment

---

[2] One exception is that service members can receive VADC and retirement benefits through the Concurrent Retirement and Disability Pay program provided that certain conditions are met. Another exception relates to surviving spouses, dependents, or children who receive indemnity compensation. The prohibition of concurrent receipt does not apply for survivors who are receiving indemnity benefits but who then qualify for a compensation or pension for their own service.

pay). The authorities usually specify the reason the armed services may use the separation benefit, the minimum eligibility criteria, and the maximum pay amount. Usually, the eligibility criteria are based on component (active component or reserve component), years of service (YOS), and retirement eligibility. The amount of the separation benefit is usually calculated as a percentage of basic pay multiplied by YOS. Separation benefits are usually paid as a lump sum, but one type of separation pay, VSI, is paid as an annuity.

The most common reasons for using these separation benefits are force shaping (e.g., reduce the size of the force) and talent management (e.g., dismiss officers who are not promoting) (DoD, 2018). Some separation benefits have been used to incentivize service members to separate voluntarily, as is the case for SSB, VSI, and voluntary separation pay (VSP). Other separation payments are involuntary separation pay (ISP), which, when used for force shaping, are akin to layoffs in civilian employment. The table also shows that readjustment pay, nondisability severance pay, and DSP are subject to recoupment by VA.

The last column in Table 2.1 shows whether the separation benefit is still in effect. Readjustment pay and nondisability severance pay ended before 1981, and SSB and VSI both ended in 2001. Note that even if services no longer use a given separation benefit, the separation benefit is still eligible for recoupment by VA.

This report focuses on a subset of the nine separation benefits: ISP, VSP, SSB, and VSI. We exclude DSP because it is part of a larger DoD disability compensation program for members found unfit to continue their military service. Those who receive a DoD disability rating below 30 percent are eligible for DSP, while those with a rating of 30 percent or above are eligible for disability retirement benefits. Note that DoD disability ratings are distinct from VA disability ratings: DoD ratings cover conditions that make a member unfit to continue to serve in the military, while VA ratings cover any condition that is service-connected. Furthermore, the congressional language specified analysis of separation pay, not severance pay. The other types of separation benefits in Table 2.1 are excluded because there are either no cases or just a handful of cases in the data we use, described in the next chapter.

TABLE 2.1
Types of Separation Pay Subject to Recoupment

Separation Benefit	Authority	Voluntary or Involuntary Separation	Minimum Eligibility Criteria	Amount of Separation Pay	Is This Still in Effect? If Not, List End Date
Readjustment pay	10 U.S.C. 687	Involuntary	Reserve officers with at least five YOS and not eligible for retirement	Lump sum calculated as 2×(monthly basic pay)×YOS; the max is the lesser of two years of basic pay or $15,000[a]	No; September 15, 1981
Readjustment pay	10 U.S.C. 3814[a]	Involuntary	Regular officers with a rank below major	Lump sum calculated as 2×(monthly basic pay)×YOS; the max is the lesser of two years of basic pay or $15,000[a]	No; December 31, 1977
Nondisability severance pay	10 U.S.C. 359, 360, 859, 860	Involuntary	On active duty on September 14, 1981; met other criteria (relating to inability to meet performance standards)	Unknown[a]	No; September 15, 1981
ISP	10 U.S.C. 1174	Involuntary	Regular and reserve officers who completed at least five years of active-duty service and do not qualify for retirement; regular enlisted service members who have completed at least six years of active-duty service and do not qualify for retirement[b]	Lump sum calculated as 0.1×12× (years of active service)×(monthly basic pay)	Yes

Table 2.1—Continued

Separation Benefit	Authority	Voluntary or Involuntary Separation	Minimum Eligibility Criteria	Amount of Separation Pay	Is This Still in Effect? If Not, List End Date
Reservists ISP (RISP)	Pub. L. 102-484, Section 4418	Involuntary	Member of the Selected Reserve with six to 15 YOS	Lump sum calculated as (YOS)×0.15×62× (daily basic pay)	Yes
SSB	10 U.S.C. 1174a (added to 10 U.S.C. 1174a by Pub. L. 102-190, Section 661)	Voluntary	Active-duty service members with six to 20 YOS (as of September 5, 1991) who then serve in the Ready Reserve and who have five or more years of continuous active-duty service immediately before separation	Lump sum calculated as 0.15×(annual basic pay)× (years of active service)	No; December 31, 2001
VSI	10 U.S.C. 1174a (added to 10 U.S.C. 1174a by Pub. L. 102-190, Section 662)	Voluntary	Active-duty service members with six to 20 YOS (as of September 5, 1991) who then serve in the Ready Reserve and who have five or more years of continuous active-duty service immediately before separation	Paid annually upon discharge; amount calculated as 0.025×(final monthly basic pay)×12×(YOS); number of payments equals 2×YOS	No; December 31, 2001
VSP	10 U.S.C. 1175a	Voluntary	Active-duty service members with six to 20 YOS; served five continuous YOS preceding receipt; not approved for VSI	Service secretary sets the amount; typically a lump-sum payment, but the service can opt for installments if the service member served 15–20 years; maximum amount is four times the comparable amount for the same grade/YOS of ISP	Yes

Table 2.1—Continued

Separation Benefit	Authority	Voluntary or Involuntary Separation	Minimum Eligibility Criteria	Amount of Separation Pay	Is This Still in Effect? If Not, List End Date
DSP	10 U.S.C. 1212	Involuntary	Have less than 20 YOS, have a disability rating less than 30 percent, and classified as unfit for duty[c]	Lump sum calculated as 2×(monthly basic pay)×(YOS up to 19)	Yes

SOURCES: 10 U.S.C. 1174; 10 U.S.C. 1174a; 10 U.S.C. 1175a; 10 U.S.C. 1212; DFAS, 2016; DoD, 2020; DoD Instruction (DoDI) 1332.29, 2017; DoDI 1332.43, 2017; Pub. L. 102-190, 1991; Pub. L. 102-484, 1992; VA, undated.

[a] The calculation varied, but, for readjustment pay, the amount in 1962 was calculated as 2×(monthly basic pay)×YOS; the max is the lesser of two years of basic pay or $15,000. However, in general, because readjustment pay and nondisability pay have long since ended, the usual sources—such as the authority itself, DoDIs, the Federal Management Regulations, and the Military Compensation Background Papers—that describe amount of severance pay are not easily accessible. As a result, we were not able to determine the amount of these separation benefits.

[b] The NDAA for FY 1991 also increased eligibility to include all members of the armed forces.

[c] Prior to January 28, 2008, service members were also required to have at least six months of service to qualify.

Recoupment

Recoupment Due to the Receipt of VA Disability Compensation

Table 2.2 shows VA recoupment amounts for each type of separation pay and specifies any exceptions to recoupment. The amount that is recouped is either the pre-tax amount of the separation benefit or the post-tax amount of the separation benefit. For most separation benefits, VA recoups the pre-tax amount if the service member received the separation benefit on or before September 30, 1996, and the post-tax amount if the service member received the separation benefit after September 30, 1996. VA recoups all separation benefits except VSI, which is instead recouped by DoD; this is a special procedure, and we explain it in the next subsection.

U.S. Department of Defense Recoupment of Voluntary Separation Incentive

The discussion above focused on VA recoupment of separation benefits whereby VA withholds a portion of VADC until the separation benefit has been recouped. In one special case, the VSI program, DoD rather than VA recoups separation benefits due to the receipt of VADC. Recall that DoD pays VSI as an annuity and not as a lump sum, unlike most other separation benefits. As a result, the process is handled by DoD. According to VA documentation,

> When a Veteran who elected to receive VSI subsequently establishes entitlement to VA compensation based on the same period of service, DoD reduces the Veteran's VSI to offset the amount of VA compensation he/she is receiving. DoD does not make deductions for VA compensation that are based on an earlier period of service than that for which VSI is payable. (VA, undated)

Because of this special case, information regarding the recoupment of VSI must come from DoD rather than VA.

TABLE 2.2

VA Recoupment of Separation Pay

Separation Pay	Recouped by VA (Yes/No)	Amount of VA Recoupment (i.e., pre-tax or post-tax amount)
Readjustment pay (10 U.S.C. 687)	Yes	• If date of entitlement to VADC is before September 15, 1981, amount of recoupment is 75 percent of pre-tax amount of the readjustment pay. • If date of entitlement to VADC is on or after September 15, 1981, and date of receipt of readjustment pay is on or before September 30, 1996, amount of recoupment is the pre-tax amount of the readjustment pay. • If date of entitlement to VADC is on or after September 15, 1981, and date of receipt of readjustment pay is after September 30, 1996, amount of recoupment is the after-tax amount of the readjustment pay.
Readjustment pay (10 U.S.C. 3814a)	Yes, in most cases	• If date of entitlement to VADC is before September 15, 1981, there is no recoupment. • If date of entitlement to VADC is on or after September 15, 1981, and date of receipt of readjustment pay is on or before September 30, 1996, amount of recoupment is the pre-tax amount of the readjustment pay. • If date of entitlement to VADC is on or after September 15, 1981, and date of receipt of readjustment pay is after September 30, 1996, amount of recoupment is the after-tax amount of the readjustment pay.
Nondisability severance pay	Yes, in most cases	• If date of entitlement to VADC is before September 15, 1981, there is no recoupment. • If date of entitlement to VADC is on or after September 15, 1981, and date of receipt of nondisability severance pay is on or before September 30, 1996, amount of recoupment is the pre-tax amount of the nondisability severance pay. • If date of entitlement to VADC is on or after September 15, 1981, and date of receipt of nondisability severance pay is after September 30, 1996, amount of recoupment is the after-tax amount of the nondisability severance pay.

Table 2.2—Continued

Separation Pay	Recouped by VA (Yes/No)	Amount of VA Recoupment (i.e., pre-tax or post-tax amount)
ISP	Yes	• If date of receipt of ISP is on or before September 30, 1996, amount of recoupment is the pre-tax amount of the ISP. • If date of receipt of ISP is after September 30, 1996, amount of recoupment is the after-tax amount of the ISP.
RISP	Yes	• If date of receipt of RISP is on or before September 30, 1996, amount of recoupment is the pre-tax amount of the RISP. • If date of receipt of RISP is after September 30, 1996, amount of recoupment is the after-tax amount of the RISP.
SSB	Yes	• Amount of recoupment is the after-tax amount of SSB.
VSI	No	• VA does not recoup VSI. DoD recoups VSI.
VSP	Yes	• If date of receipt of VSP is on or before September 30, 1996, amount of recoupment is the pre-tax amount of the VSP. • If date of receipt of VSP is after September 30, 1996, amount of recoupment is the after-tax amount of the VSP.
DSP	Yes, in most cases[a]	• If date of receipt of DSP is on or before September 30, 1996, amount of recoupment is the pre-tax amount of the DSP. • If date of receipt of DSP is after September 30, 1996, amount of recoupment is the after-tax amount of the DSP.

SOURCE: VA, undated.

[a] DSP is not recouped if (1) the service member separated on or after January 28, 2008, and (2) the disability for which the member received the DSP was incurred in the line of duty in a combat zone or during combat-related operations.

Recoupment Due to the Receipt of Military Retired Pay

As mentioned in Chapter One, veterans may also experience the recoupment of separation pay if they qualify for military retired pay, even if they do not qualify for VADC.[3] That said, because of our study's focus on separation pay recoupment due to the receipt of VADC, and because some veterans also qualify for military retired pay, we also consider recoupment from military retired pay. (That is, we consider recoupment of the overlap area of the two circles in Figure 1.1). For example, consider a service member who separated and took VSI at the time of separation. Members who took VSI were required to remain in the Ready Reserve of a reserve component for at least three years, so this service member could later qualify for retirement either based on service in the reserves, or, if the service member was called up, based on time in active duty. This member could also experience a service-connected disability related to their time in service before separation that would qualify for benefits from VA. In this case, the member could eventually receive both VADC and military retired pay, and the member's VSI payments would be recouped from both.

VA and DoD must coordinate their recoupment efforts in these cases. In particular, VA first deducts separation pay from VADC. This often occurs before the member qualifies for retirement. Once a member qualifies for retirement, the remaining separation pay can then be recouped by the DoD from military retired pay at a rate of 40 percent of monthly retired pay.

---

[3]   This is done to avoid paying the service member twice based on their service. It is also important to note that this is done in such a way as to avoid the imposition of undue financial hardship. Specifically, 10 U.S.C. 1174(h) states,

> Coordination with Retired or Retainer Pay and Disability Compensation.—(1) A member who has received separation pay under this section, or separation pay, severance pay, or readjustment pay under any other provision of law, based on service in the armed forces, and who later qualifies for retired or retainer pay under this title or title 14 shall have deducted from each payment of such retired or retainer pay an amount, in such schedule of monthly installments as the Secretary of Defense shall specify, taking into account the financial ability of the member to pay and avoiding the imposition of undue financial hardship on the member and member's dependents, until the total amount deducted is equal to the total amount of separation pay, severance pay, and readjustment pay so paid.

Trends in Separations with Involuntary Separation Pay, Voluntary Separation Incentive, and Special Separation Benefits

Not all separations from military service occur with the payment of a separation benefit, and not all recipients of a separation benefit will qualify for VADC and therefore be subject to recoupment. In our study, we sought to estimate the number of individuals subject to separation pay recoupment due to the receipt of VADC. To provide contextual background on the population who could be subject to recoupment, we show the trends in the number of recipients of ISP, VSI, and SSB.

Figure 2.1 shows the trend over time in separations with nondisability separation pay under 10 U.S.C. 1174, or ISP since 1981. ISP separations increased during the post–Cold War drawdown years in the early and mid-1990s from 2,333 in 1989 to 12,978 in 1997. They fell in the early 2000s to a low of 3,450 in 2002. Involuntary separations spiked in 2008 to over 26,800 and have been decreasing since 2014.

Table 2.3 shows the number of separations with SSB and VSI for 1992 to 2001. As mentioned earlier, SSB and VSI were used to reduce the size of

FIGURE 2.1

Trend over Time in Number of Separating Personnel Receiving ISP

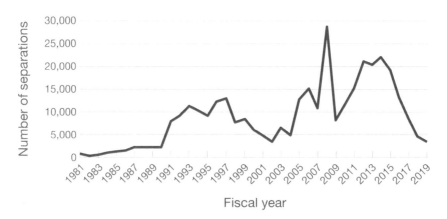

SOURCES: DoD, 2018, for 1981 to 2015; DoD budget materials for 2016 to 2019 (see Office of the Under Secretary of Defense [Comptroller], undated).

TABLE 2.3

Number of Separations with SSB or VSI, 1992 to 2001

Fiscal Year	SSB	VSI
1992	56,475	9,659
1993	10,593	7,120
1994	9,271	9,528
1995	7,455	8,891
1996	413	6,209
1997	167	1,527
1998	806	945
1999	38	165
2000	35	162
2001	44	79

SOURCE: Data from 1992 to 2020 provided to the authors by DFAS.

the military following the end of the Cold War. The table shows that VSI (the annuity option) was less common among those separating with these benefits than SSB (the lump-sum option). In total, more than 66,000 service members separated with either benefit in 1992, shrinking to 123 in 2001.[4]

Figure 2.1 and Table 2.3 show trends in separations by year. But a veteran who received separation pay in any past year could qualify for VADC or military retired pay, or both, at any point and therefore be subject to recoupment. Put differently, data on veterans subject to recoupment at a point in time reflect the total of all past recipients of separation benefits who qualified for VADC or military retired pay, or both, and who are in the process of having their separation pay recouped. We discuss the data for our analysis in the next chapter.

---

[4] The fact that so many members selected the lump-sum option over the annuity suggests that these members have high discount rates—a point analyzed in Warner and Pleeter, 2001.

CHAPTER THREE

Data and Research Approach

We used several administrative data sources for this project that were provided by DFAS and VA. The data include information about the amounts of recoupment from VA disability benefits and the number of service members affected by recoupment due to receipt of VA disability benefits and recoupment due to receipt of military retired pay. We describe all of the data sources in the following sections.

Data on Recoupment from VA Disability Compensation

Special Separation Benefits and Separation Pay

VA is the agency in charge of recoupment of SSB and separation pay from VADC and keeps the administrative records for these recoupment activities. For this project, VA provided records with any recoupment of separation pay or SSB from VADC from FY 2013 until the time the data were extracted for our project in January 2021. The data fields for recoupment of separation pay include recoupment from either VSP or ISP. The records also provide data on the full amount of VSP or ISP or the amount of SSB paid and the full amount recouped for any veteran who had any recoupment occur during this time. Furthermore, the records include the gross and after-tax amounts of separation pay and the date on which VA benefits began. For cases in which recoupment is ongoing, we also observe the amount recouped to date and any amount of recoupment outstanding. Using these records, we can calculate the following for VSP/ISP and SSB:

1. the total number of service members who experienced recoupment of VA benefits from VSP/ISP or SSB since 2013
2. the total number of service members with ongoing recoupment of VA benefits from VSP/ISP or SSB as of January 2021
3. the total amount of separation pay recouped to date from cases in #1 and #2
4. the total amount of outstanding recoupment for cases in #2.

VA underwent a significant update to its data entry systems in FY 2013. Because of changes in the structure of the administrative records, we do not have data on cases in which recoupment was completed prior to FY 2013.

Voluntary Separation Incentive

Recall that DoD reduces the annual VSI payments by the amount of VADC received in a given year to account for recoupment of VA benefits. As a result, recoupment of VSI due to receipt of VA benefits is handled by DFAS. DFAS provided two databases related to recoupment of VSI. The master record for all VSI payments is the 2212 file, a database that provides information on the universe of all VSI payments ever awarded between 1992 and 2001. This data file includes details about the separation date, the first and last date of VSI payment (including expected dates for cases in which VSI benefits are still being paid), the total gross amount of VSI per person, and the amount of VSI paid to date as of the time the data were extracted in December 2020. For cases in which VSI payments are complete, we can infer the amount of recoupment of VA benefits as the difference between the total gross amount of VSI and the amount paid to the service member, if this difference is positive.

Next, DFAS provided annual records of payment for VSI for service members who are still receiving VSI in 2020. These records provide information about the annual amount of VSI paid, annual deductions due to recoupment of VA benefits, and annual deductions for federal and state taxes. Using these annual records, we observe the amount of annual recoupment for service members who were still receiving VSI at the end of 2020.

We can also link the total gross VSI award and expected last date of VSI payment from the annual files to information in the 2212 file.

Because we observe the annual amount of recoupment as well as the number of years of VSI payments remaining, we multiply these two values to estimate the total amount of recoupment outstanding for service members for whom recoupment is ongoing. This estimation requires the assumption that the amount of VADC will remain the same for all future years of recoupment.

However, we are not able to observe or estimate the total amount recouped prior to 2020 for cases in which VSI payment and recoupment are ongoing. As explained above, we infer the total amount recouped for completed cases based on the difference between the total gross amount of VSI and the amount paid to date. For cases with ongoing VSI, the difference between the total gross amount and the amount paid to date reflects any outstanding annual payments to be made as well as any amount recouped to date. If we were able to observe the date on which VA benefits began, we could provide estimates that would disentangle these two amounts. However, the first date of VA benefits is not included in the DFAS administrative data, meaning that we cannot distinguish which portion of the outstanding value can be attributed to recoupment rather than outstanding payments. Figure 3.1 summarizes the periods of recoupment of VSI that we observe for veterans with ongoing and completed VSI payments.

FIGURE 3.1

Observed and Estimated Recoupment of VSI for Completed and Ongoing Payments

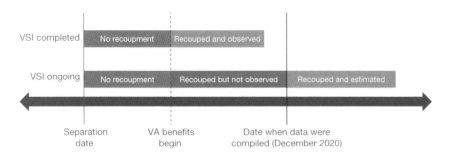

In summary, we can quantify the following for VSI:

1. the total number of service members with completed recoupment of VA benefits from VSI
2. the total number of service members with ongoing recoupment of VA benefits from VSI as of December 2020
3. the total amount of VA benefits recouped from VSI for service members who have completed recoupment
4. an estimate of the total amount of VA benefits recouped from VSI among cases with ongoing recoupment.

Tables 3.1 and 3.2 summarize the information we observe about recoupment of VA benefits from each type of separation pay, depending on the data source.

As these tables show, and as described earlier, the structure of the various data sources means that our analyses cover different periods depending on the type of separation pay. In the case of VSI, the data structure also provides different information for cases in which the VSI annuity has been fully paid, compared with cases in which annual payments are ongoing.

TABLE 3.1

Type of Separation Pay and Observed Recoupment in Each Data Source

Separation Pay Type	Date Range of Eligibility	Data Source	Completed Recoupment Observed in Data	Ongoing Recoupment Observed in Data
VSP or ISP 10 U.S.C. 1174 and 10 U.S.C. 1175a	1991–present	VA	Cases with any recoupment occurring in 2013 or later	All ongoing as of January 2021
SSB (10 U.S.C. 1174a)	1992–2001	VA	Cases with any recoupment occurring in 2013 or later	All ongoing as of January 2021
VSI (10 U.S.C. 1174a)	1992–2001	DFAS	1992–2020	All ongoing as of December 2020

TABLE 3.2

Information Available for Computing Recoupment for Each Type of Separation Pay

Information component	VSI (ongoing)	VSI (completed)	SSB	VSP or ISP
Start/end dates of payment	X	X		
Start date of VA benefits			X	X
Amount of total award	X	X	X	X
Amount recouped to date		X	X	X
Amount outstanding to be recouped	X		X	X
Annual amounts	X			

Data on Recoupment from Military Retired Pay

DFAS also provided annual records of cases in which VSI, SSB, or other separation payments are being recouped from military retired pay. These cases correspond to the blue circle in Figure 1.1. DFAS provided these records for 2013 to 2020, matching the time frame over which we have data on recoupment of VADC from VA. These records provide information on the type of separation pay being recouped, the total amount that has already been recouped from VA (if any), the total amount recouped to date from military retired pay (as of December in the year when the data were pulled), the total remaining balance to be recouped, and the monthly rate of recoupment from military retired pay. The data indicate whether the separation pay award was SSB, VSI, or "other" separation payments, where "other" includes ISP, VSP, nondisability separation pay, or DSP. Most separation payments in the "other" category are either ISP or VSP.[1] We can derive the total separation pay award for VSI and for SSB by using information from the 2212 file

---

[1] The data also include a field for recoupment of readjustment pay, but these cases are extremely rare, and we exclude them from our analysis. Nondisability severance pay has been superseded by involuntary separation, per DoD 7000.14-R, Vol. 7B, Chapter 4 (DoD, 2021b). Recoupment of DSP is included in the database for cases in which a service member's case was sent to the Physical Disability Board of Review (PDBR) and the individual received an increase in their disability rating that made them eligible for

on VSI and a similar file with a list of all SSB payments (called the 2213 file). For other types of separation pay, the total award value is typically included in the recoupment data file directly.[2]

Using these annual files, we estimate the following components related to recoupment from military retired pay:

1. the number of service members with any recoupment of separation pay from military retired pay between 2013 and 2020
2. the total amount of separation pay recouped between 2013 and 2020 (per person, and in total)
3. the number of service members with recoupment from both military retired pay and VA disability benefits between 2013 and 2020 (i.e., the overlapping part of the two circles shown in Figure 1.1).
4. the amount recouped from VA benefits for members experiencing recoupment from both military retired pay and VA disability benefits between 2013 and 2020.

Research Approach

Section 654 of the NDAA for FY 2020 requires that the study of recoupment provides information on the number of service members and veterans who receive separation pay who are affected by recoupment and the aggregate amount of money they would receive in the absence of recoupment (Pub. L. 116-92, 2019). We conduct a series of tabulations to respond to this request. We quantify the total number of members and veterans who

---

disability retirement pay. We also exclude DSP in our analyses to focus on VSP and ISP, which are what are included in the data from VA.

[2] The recoupment administrative file restricts each record to a maximum of $99,999 per entry, and entries in which recoupment has been completed are removed from the database every six months. So, we cannot observe the full separation payment amount for cases in which total separation pay exceeded $99,999 if that portion of recoupment was completed more than six months before the data were extracted. However, over 95 percent of payment amounts for separation pay observed in the data are less than $99,999, meaning that we can observe the total separation pay amount in nearly all cases.

are subject to recoupment of separation pay from VA benefits and from military retired pay, subject to the year limitations described above. We then estimate the amount that would have been received in the absence of recoupment in two ways. First, we compute the average amount of recoupment each member has experienced. Second, we estimate the aggregate total amount of recoupment when summed across all service members. We calculate these numbers separately for members who have completed recoupment and for members for whom recoupment is ongoing.

There are some limitations to what we can estimate. First, because we do not observe the start date for VA benefits for cases in which there was recoupment from VSI, we cannot estimate the total amount recouped to date for cases in which VSI is ongoing. We also cannot quantify the entire population of members who ever experienced recoupment of SSB or separation pay before FY 2013. Second, because we do not know the entire population of members who ever received separation pay or the dates of separation for the cases with recoupment of VSP or ISP, we cannot project into the future the share of members who could be subject to recoupment in the future. And finally, we observe only cases in which VADC was recouped, but not the total number of veterans who received VADC. As a result, we cannot project how many service members who receive VADC will ever experience recoupment because of receipt of separation pay.

CHAPTER FOUR

Results

In this chapter, we present our findings in response to both questions in the congressional mandate:

1. How many members and veterans are affected by recoupment of separation pay (including separation pay, SSB, and VSI) due to receipt of VA benefits?
2. What is the aggregated amount of additional money members and veterans would receive in the absence of such recoupment?

In addition, we present similar findings with respect to the recoupment of separation pay from military retirement pay.

Question 1: How Many Members and Veterans Are Affected by Recoupment of Separation Pay?

Figure 4.1 presents the total number of veterans who have been subject to recoupment of separation pay (VSP or ISP) or SSB since FY 2013. In total, 72,206 veterans have been subject to recoupment of VSP or ISP between 2013 and 2020. Among those veterans, 58,542 have completed recoupment of the separation pay, while recoupment was ongoing for the remaining 13,664 veterans at the time our data were collected in January 2021.

Substantially fewer veterans were subject to recoupment of SSB during this period. A total of 2,651 veterans were subject to recoupment of SSB sometime between 2013 and 2020, and only 277 veterans had ongoing recoupment at the time the data were collected. Veterans with ongoing recoupment were likely rated within the past few years and recently began

FIGURE 4.1

Count of Veterans with Any Recoupment of Separation Pay (VSP or ISP) or SSB Due to Receipt of VADC Between 2013 and 2020

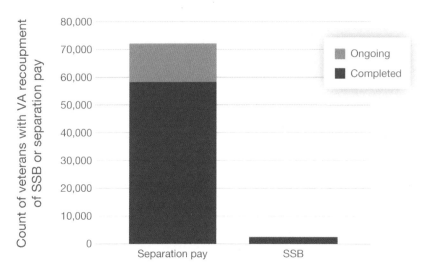

SOURCE: Data provided to the authors by VA.
NOTE: Sample includes any veteran with any recoupment of SSB or separation pay that occurred sometime between FY 2013 and FY 2020. The label "Separation pay" refers to VSP or ISP.

receiving VADC. Recall that, for cases included in the data, we do not observe recoupment of VSP/ISP or SSB that occurred and was completed before 2013 or the dates of separation. As a result, we are unable to compute the share of separation pay awarded or the total share of all SSB awards that were ever subject to recoupment.

Figure 4.2 shows the number of veterans who experienced recoupment of VSI since the program was enacted in 1992. In total, 4,701 veterans have experienced recoupment of VSI. This total represents 10.6 percent of all VSI awards. At the time the data were collected in December 2020, 1,405 veterans had ongoing recoupment from VSI, while 3,296 veterans had competed their recoupment of VSI.

Next, we estimate when recoupment from VSI will be completed for the cases in which it is ongoing. This estimate provides information on how soon recoupment will end for the VSI program. To make this estimate, we assume that recoupment will continue to occur in each of the remaining

FIGURE 4.2

Count of Veterans with Recoupment of VSI Due to Receipt of VADC

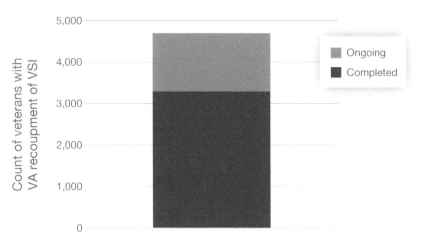

SOURCE: Data from 1992 to 2020 provided to the authors by DFAS.
NOTE: The recoupment totals include cases of receipt of both VA benefits and military retired pay but exclude cases of recoupment based on receipt of military retired pay only.

years of payment for the VSI annuity.[1] The top panel in Figure 4.3 shows the number of veterans currently experiencing recoupment who will complete VSI and recoupment in each of the next 20 years.[2] The bottom panel shows the cumulative share of all veterans who will complete recoupment of VSI in each of the next 20 years. In 2021, 225 veterans with ongoing recoupment of VSI (approximately 20 percent) will receive their final VSI payment and, thus, the final recoupment due to receipt of VA benefits. This share increases sharply over time. Within the next five years (i.e., by the end of 2025), two-thirds of all cases with outstanding recoupment of VSI due to

---

[1] It is possible for members to terminate their outstanding VSI payments, meaning that this assumption yields an upper bound on the number of years of recoupment remaining and an upper bound on the last year in which VSI payment and recoupment would occur for each case with outstanding recoupment.

[2] Note that veterans currently receiving VSI might become eligible for VA benefits in the future and thus start to experience recoupment, so these numbers might change as time progresses.

FIGURE 4.3

Count and Cumulative Share of Veterans with Outstanding VSI Recoupment, Based on Completion Year of VSI and Recoupment

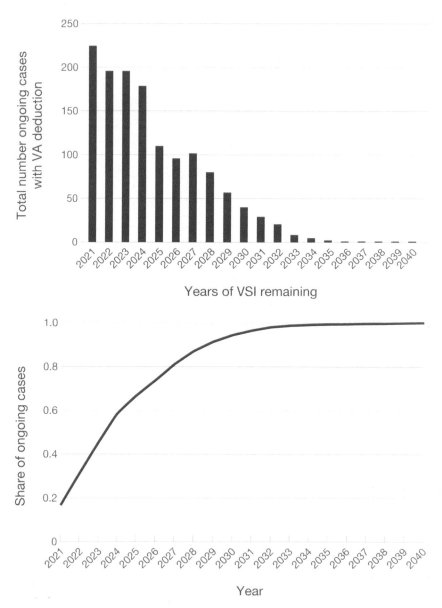

SOURCE: Data from 1992 to 2020 provided to the authors by DFAS.

NOTE: The recoupment totals include cases of receipt of both VA benefits and military retired pay but exclude cases of recoupment based on receipt of military retired pay only.

receipt of VA benefits will be completed. There are some cases in which VSI payments will continue until as late as 2040, although these cases represent a minority of the remaining VSI population. In short, by 2035, virtually all currently ongoing recoupment cases of VSI will be completed.

In sum, several key takeaways emerge regarding the first question posed by Congress. First, in recent years, the number of veterans experiencing VA recoupment of separation pay is much larger than the number experiencing VA recoupment of SSB: about 72,000 versus 2,651 between 2013 and 2020. It is also larger than the number experiencing recoupment of VA benefits from VSI. Since 1992, we find that about 4,700 veterans have experienced recoupment from VSI. Second, only a small share (10.6 percent) of VSI awards have ever been subject to recoupment of VA benefits. Finally, we find that about two-thirds of outstanding VSI recoupment cases will be completed within the next five years.

Question 2: What Is the Aggregated Amount of Additional Money Members and Veterans Would Receive in the Absence of Recoupment?

For each type of separation pay, we estimate the aggregated amount of additional money that veterans would receive but for recoupment. We present these calculations in two ways: First, we show average per-person total amounts, and then we present aggregated totals summing the per-person amounts across all individuals who have experienced or are experiencing recoupment. The first way shows the computation from the perspective of the individual veteran, while the second way shows the computation from the perspective of the government or agency. Because the nominal value of separation pay is the amount that is recouped, all of our calculations are in nominal terms.

Before presenting these calculations for each of the pay types, we provide context about the overall value of the various types of separation pay to service members. Table 4.1 shows the median amounts and the range of amounts of SSB, VSI, and ISP/VSP for the years observed in our data set. In the first two columns, we show these amounts including all records in the data, regardless of whether there was any recoupment. The median value

TABLE 4.1

Amounts and Distribution of Separation Pay

Separation Pay Type	Median Value of Total Award Overall	Total Award Range: 5th to 95th Percentile	Median Value of Total Award Among Cases with Recoupment
VSP or ISP	$17,700	$2,000–$59,200	$17,700
SSB	$31,700	$11,000–$63,100	$25,600
VSI	$16,200	$2,500–$400,000	$175,400 (VSI complete) $342,000 (VSI ongoing)

SOURCES: Data from 1992 to 2020 provided to the authors by DFAS; data from 2013 to 2020 provided to the authors by VA.
NOTES: VSI computation for 1992 to 2001; SSB overall total for 1992 to 2001; separation pay and SSB recoupment totals for 2013 to 2020. Values shown in nominal dollars. Data provide information only on the value of VSP or ISP for cases with recoupment.

of the VSI award is in a similar range and, in fact, slightly lower than the median values for ISP/VSP and SSB. However, VSI awards have a significantly larger range compared with the other separation payments, ranging from a low of $2,500 to a high of over $400,000. One possible explanation for this range could be that service members who selected VSI had differentially longer careers or higher basic pay at the time of separation, leading to significantly larger total awards.[3]

In the rightmost column, we show the median value of awards that were subject to recoupment. As shown, the VSI awards experiencing recoupment tend to be at the upper end of the range shown in the middle column, significantly larger than the average VSI award. Although we do not have sufficient information to precisely explain the reason why the awards experiencing recoupment tend to be so much larger, one possible explanation could be mechanical. Because larger awards often are paid for a longer period, there is a longer time frame during which the veteran could also qualify for VADC and thus be subject to recoupment. Other possible reasons could be related to different characteristics of the veterans who experience recoup-

---

[3] However, we do not observe monthly basic pay or YOS in the recoupment data provided, which would allow us to test these possible explanations. That said, Table 1 of Warner and Pleeter, 2001, indicates that the percentage of personnel in their data selecting VSI over SSB increased as grade and YOS increased.

ment and those who do not, such as their pay grade, service, and occupation. An exploration of these possible explanations was beyond the scope of our study.

Separation Pay (Voluntary Separation Pay or Involuntary Separation Pay) and Special Separation Benefits

Figure 4.4 shows the average per-person after-tax value of separation pay and SSB among awards that were subject to recoupment between FY 2013 and FY 2020. These values reflect the full amount that was subject to recoupment of separation pay (VSP or ISP) or SSB. Put differently, the amount of VADC a veteran would have received but for recoupment equals the total amount of after-tax separation pay or SSB award, and the averages of these amounts are shown in Figure 4.4.

FIGURE 4.4

Average Per-Person After-Tax Value of Separation Pay (VSP or ISP) and SSB Among Awards Subject to Recoupment, 2013–2020

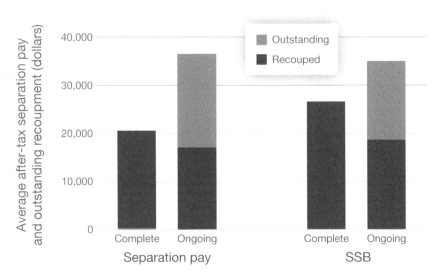

SOURCE: Data from 2013 to 2020 provided to the authors by VA.

On average, the after-tax value of VSP or ISP among cases in which recoupment was completed between 2013 and 2020 was just over $20,000. The average after-tax amount of SSB among cases in which recoupment was completed between 2013 and 2020 was approximately $26,000. Veterans with ongoing recoupment had larger amounts of separation pay and SSB awards; this is likely the case because VADC must be withheld for a longer period to recoup a larger amount. The average per-person after-tax value among cases in which there is ongoing recoupment is approximately $36,000 for cases with recoupment of separation pay and $35,000 for cases with recoupment of SSB. On average, approximately half of these amounts have been recouped to date.

Next, we sum the total amount of separation pay and SSB subject to recoupment across all veterans who experienced recoupment of either pay type between 2013 and 2020.

Separation Pay (Involuntary Separation Pay or Voluntary Separation Pay)

In total, approximately $1.4 billion has been withheld from VADC because of the receipt of ISP or VSP since 2013. In Figure 4.5, the $1.4 billion is the sum of the two blue bars. Approximately $1.2 billion of this amount came from cases in which recoupment has been completed, and the remaining $233 million is the amount recouped to date from cases in which recoupment is ongoing. The amount of outstanding recoupment of separation pay due to receipt of VADC as of January 2021 was approximately $264 million (the green bar). Because eligibility for VSP and ISP is ongoing, new veterans likely will become eligible for VADC and be subject to recoupment on an ongoing basis, potentially adding to this outstanding amount to be recouped.[4]

The tabulations in Figure 4.5 show aggregate recoupment of VSP or ISP since 2013. We estimate the annual rate of recoupment from VADC due

---

[4] Recall that, while we observe the date on which VA compensation began, we do not observe the date of separation. As a result, we cannot compute the share of separations in a given year that have been subject to recoupment in order to project the share likely to be subject to recoupment on an ongoing basis in the future or to infer whether the outstanding value of recoupment represents a steady state.

FIGURE 4.5

Aggregate Total Amount of VSP or ISP Recouped, 2013–2020

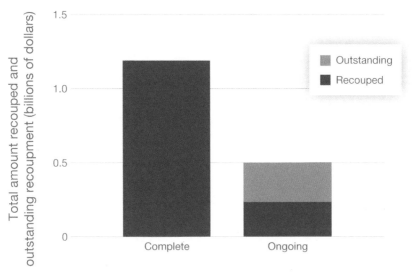

SOURCE: Data from 2013 to 2020 provided to the authors by VA.

to the receipt of VSP or ISP by dividing the total amount recouped to date by the number of years since the veteran began receiving VADC, assuming that the amount of VADC was constant over this time. For cases with outstanding recoupment of separation pay, we estimate $4,900 recouped from the veteran annually. If we take the average award values as shown in Figure 4.4, this means that it would take approximately seven years to fully recoup the average separation pay award of $35,000.[5] Under the assumption that an individual's VADC amount remains constant, this per-person amount implies that an aggregate amount of approximately $67 million will be recouped because of the receipt of VSP or ISP within the next year.[6]

---

[5] The data do not provide an end date for completed cases, so we are unable to estimate an annual amount of recoupment for completed cases. If we assume that the annual rate of $4,900 is similar for completed cases, this means that recoupment took approximately five years, on average, for the completed cases observed in the data.

[6] This estimate, and the annual estimate for SSB, is based on the cases for which we are able to observe recoupment. However, as mentioned earlier, it is possible that

Special Separation Benefits

Figure 4.6 shows the aggregated total amount recouped from VADC between 2013 and 2020 for SSB. Although the per-person values were roughly similar for SSB and VSP/ISP in Figure 4.4, the aggregated total is much lower for SSB because fewer veterans experienced recoupment of SSB. In total, approximately $68 million was recouped because of the receipt of SSB since FY 2013 (the sum of the two blue bars in Figure 4.6), with an outstanding balance of approximately $5 million (the green bar). Using the same calculation as above to estimate the annual amounts of recoupment, we estimate that the average annual per-person amount of recoupment due to receipt of SSB is $2,600, implying that approximately $730,000 in aggregate will be recouped in the next year.

FIGURE 4.6

Aggregate Total Amount of SSB Recouped, 2013–2020

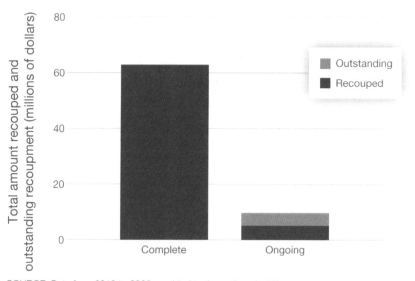

SOURCE: Data from 2013 to 2020 provided to the authors by VA.

more service members will qualify for VADC in the future and will begin experiencing recoupment.

Voluntary Separation Incentive

As explained earlier, both the structure of VSI payments and the process for recoupment from VSI differ from those of the other types of separation pay. First, VSI is paid as an annuity for two times the number of YOS at the time of separation. Second, DoD is responsible for recoupment of VSI; DFAS reduces these annuity payments by the amount of VA compensation received in a given year. Furthermore, recall that our data source for VSI provides different information about the amount of recoupment depending on whether the VSI annuity has been completed or is ongoing. For cases in which the annuity has been completed, we observe the full amount recouped. For cases in which the annuity is ongoing, we observe the annual amount withheld from the annuity. For all cases, we observe the first and last dates of VSI payment. With this information, we can project the total value of outstanding recoupment for cases in which the annuity and, thus, recoupment are ongoing, under the assumptions that VADC remains constant and that the veteran does not elect to terminate the remaining VSI payments. Another way in which VSI differs from the other separation pay programs is that the VSI awards have a significantly larger range. These differences provide context for understanding the recoupment totals shown below.

First, we focus on the 3,296 cases in which recoupment of VSI has been completed and we observe the full value that was recouped. To recoup VSI, DoD reduces annual VSI payments by the amount of VADC received in the year. In nearly all cases, this reduction is less than the full value of VSI received that year, meaning that the veteran receives a reduced payment but does not have their full VSI amount withheld. As shown in Table 4.2, the average (median) total award was $190,000 ($175,000) and the average (median) amount of VSI recouped was $53,000 ($36,000). On average, 27 percent (or about $53,000/$190,000) of an individual's total VSI was recouped because of receipt of VADC, among cases in which recoupment was completed. When we aggregate over all completed awards, the total amount of recoupment from completed VSI amounts to $177 million.

While we are unable to observe the total amount recouped to date from the 1,405 cases in which VSI recoupment is ongoing, we can estimate the total amount of VSI that is outstanding for these cases. On average, approximately $5,000 is recouped each year from VSI annuities among cases in which recoupment is outstanding. Because SSB, VSP, and ISP are lump-

TABLE 4.2

VSI Award and Recoupment Totals Among Cases with Completed Recoupment

Metric	Total VSI Award	Amount Recouped
Individual average (mean)	$190,000	$53,000
Individual median	$175,000	$36,000
Aggregate total amount recouped	—	$177 million

SOURCE: Data from 1992 to 2020 provided to the authors by DFAS.
NOTE: The dash indicates that the aggregate total amount could not be calculated due to data limitations.

sum payments, they are recouped retroactively from VADC when the veteran later begins receiving VA benefits. By contrast, VSI is an annuity and is recouped only while the annuity payments are ongoing. So, the number of years that the veteran is subject to recoupment of VSI depends on when the veteran begins receiving VADC relative to how many years are left in the annuity. Figure 4.7 shows the average total outstanding amount of VSI that will be recouped in each of the next 20 years and the cumulative share of the amounts to be recouped by year. We estimate that two-thirds of the total amount of outstanding VSI recoupment will be completed in the next five years, consistent with our finding in Figure 4.3 showing the number of people who will complete VSI in each of these years. We estimate that approximately $9.6 million will be recouped from VSI in 2021.

Several key takeaways also emerge with respect to the second question posed by Congress. First, we find that SSB and VSI awards for which recoupment has occurred are significantly larger than the average SSB and VSI awards. That is, individuals who experience recoupment have larger SSB and VSI awards. Furthermore, the average amount of recoupment per person is similar for veterans who received SSB or other separation pay (ISP or VSP), while the average amounts recouped from veterans who received VSI are higher. That said, the aggregated total amount recouped across all cases is smaller for VSI and SSB compared with VSP/ISP because the number of VSP/ISP cases far exceeds the number of SSB or VSI cases. Second, unlike SSB and separation pay, the full amount of which is recouped by VA, only a portion of VSI is typically recouped by DFAS because the annual amount of

FIGURE 4.7

Annual Amount Expected to be Recouped from VSI by Year, and Cumulative Share of Outstanding Amounts

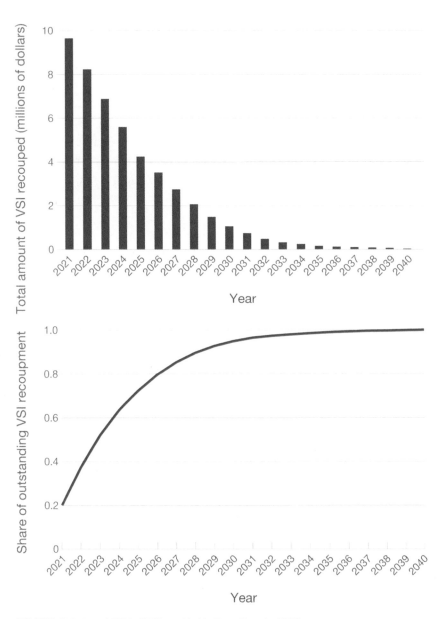

SOURCE: Data from 1992 to 2020 provided to the authors by DFAS.

VA compensation is less than the VSI amount. Finally, the majority of veterans who have been subject to recoupment over the past seven years—and the majority of veterans who still are experiencing recoupment—received either VSP or ISP, not SSB or VSI. Looking ahead, the main type of separation pay that will be subject to recoupment will be VSP or ISP because cases of VSI and SSB will eventually decline to zero given that these programs ended in 2001.[7]

Military Retired Pay

DFAS also provided data on recoupment of separation pay from military retired pay as part of our data request for this project, so we can calculate the number of veterans who experience recoupment of separation pay because of receipt of military retired pay and the amount of additional money veterans would receive in the absence of recoupment of separation pay from retired pay.

Figure 4.8 shows the number of veterans experiencing recoupment of VSP/ISP, SSB or VSI from military retired pay between 2013 and 2020. In total, we observe approximately 14,900 veterans with recoupment from military retired pay—a much lower total than the number of veterans experiencing recoupment from VADC over the same period. This number is the

---

[7] Note that this conclusion is based on cases with ongoing recoupment observed at the time the data were collected. Because VSP and ISP are continuing programs, service members will continue to separate with separation pay (i.e., VSP or ISP) and become subject to recoupment in the future. Although current service members are not able to receive VSI or SSB, it is also possible that some veterans who received these payments in the past may begin receiving VADC in the future and thus become subject to recoupment in the future. However, we suspect this scenario to be uncommon for several reasons. First, SSB and VSI payments were available only for service members who separated between 1992 and 2001, meaning that 20 years have now passed since the last time a member qualified for either payment. Thus, the majority of VSI or SSB recipients who will qualify for VADC likely have already done so. Secondly, VSI recoupment of VADC occurs only while a member is receiving ongoing annuity payments. However, the majority of VSI payments are complete. According to the last dates of payment in the file, only 2,325 members who received VSI (approximately 5 percent of all VSI awards) have not yet experienced recoupment and have outstanding payments to be made. The dates of last payment imply that 70 percent of these payments will be completed in the next five years.

FIGURE 4.8

Count of Veterans with Any Recoupment of Separation Pay, SSB, or VSI Due to Receipt of Military Retired Pay Between 2013 and 2020

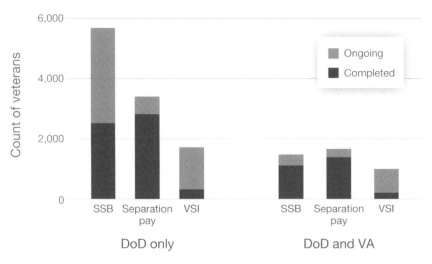

SOURCE: Data from 2013 to 2020 provided to the authors by DFAS.
NOTE: The label "Separation pay" refers to VSP or ISP.

sum of all columns in Figure 4.8 and reflects the number of veterans in the full blue circle in Figure 1.1. Figure 4.8 separates this total by pay type, and by whether the veterans are experiencing recoupment only from retired pay (DoD only), or whether they experience recoupments from military retired pay and VADC. Recall that the latter group is shown in the intersection of the two circles in Figure 1.1.

Approximately 10,800 veterans experienced recoupment only from retired pay between 2013 and 2020 (i.e., the sum of the columns labeled "DoD only" is 10,800). Approximately 5,700 veterans experienced recoupment of SSB, 3,400 experienced recoupment of VSP or ISP, and 1,700 experienced recoupment from VSI. The three bars on the right side of the graph show that just over 4,000 veterans experienced recoupment from both retired pay and VADC. Among this group, nearly 1,500 experienced recoupment of SSB, more than 1,600 experienced recoupment of VSP or ISP, and just under

1,000 experienced recoupment from VSI. In total, we observe nearly 6,500 cases with ongoing recoupment from military retired pay in 2020.

While the number of veterans experiencing recoupment of VSI from military retired pay is in a similar ballpark as the number experiencing recoupment from VADC, the number of veterans experiencing recoupment of SSB from retired pay is higher than the number of veterans experiencing recoupment of SSB from VADC. Because veterans must reach age 60 to begin receiving military retirement benefits, it is possible that the counts are higher for SSB because veterans who took SSB in the late 1990s are now reaching retirement age. Indeed, Figure A.1 in the appendix shows that the number of veterans with recoupment of SSB has increased over the past five years. By contrast, the number of veterans experiencing recoupment of VSP or ISP from military retired pay is much lower than the number of veterans experiencing recoupment of separation pay from VADC. It is possible that relatively few individuals who separate and receive VSP or ISP are eligible for retirement.

Figure 4.9 shows the average amount recouped to date for cases with completed and ongoing recoupment and the average amount outstanding for cases with ongoing recoupment.[8] The chart on the left shows the averages for cases with completed recoupment, and the chart on the right shows the averages for cases in which there was an outstanding balance (in other words, recoupment was ongoing) at the end of 2020. In each chart, the blue section of the bars reflects the amount recouped from retired pay, the orange section shows the amount recouped from VADC, the green section shows the amount by which VSI was reduced to account for VADC recoupment, and the purple section shows the amount remaining to be recouped from retired pay (i.e., the amount outstanding at the end of 2020).[9]

---

[8] We characterize cases as completed if we observe the end of the recoupment between 2013 and 2020, even if recoupment began prior to 2013. The structure of the data typically allows us to observe the full amounts recouped for SSB and VSP/ISP, although the information on recoupment prior to VSI can be incomplete because of the fact that the database erases subrecords of completed recoupment that exceeded $99,999. Figure A.2 in the appendix shows that the average amount of awards is similar among cases for which we observe recoupment from beginning to end between 2013 and 2020.

[9] Our data on recoupment from military retired pay do not provide information on the amount outstanding to be recouped from VADC (if any).

FIGURE 4.9

Average Per-Person Amount Recouped to Date from Separation Pay, SSB, and VSI Among Awards with Ongoing Recoupment from Military Retired Pay

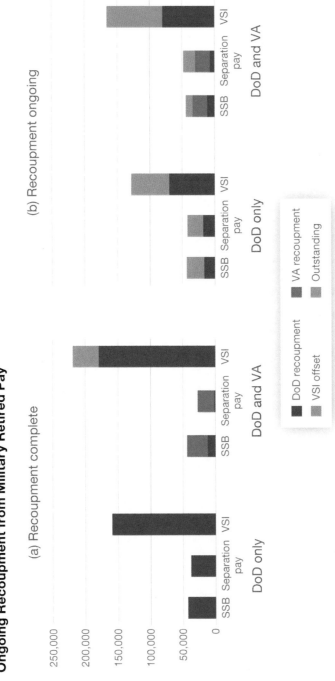

SOURCE: Data from 2020 provided to the authors by DFAS.

NOTE: Estimate of VSI offset from VA is based on recoupment of VSI among cases in which VSI annuity is complete, which account for approximately two-thirds of all cases with dual recoupment of VSI from DoD and VA. The label "Separation pay" refers to VSP or ISP.

We begin by focusing on the chart on the left, which shows the values among cases in which recoupment is complete. The average SSB award was $43,000 among cases with recoupment only from retired pay (i.e., the "DoD only" cases), and the average VSP or ISP awards were approximately $38,000 among completed cases with recoupment only from retired pay. Among completed cases with recoupment from both retired pay and VADC (i.e., the "DoD and VA" cases), the average SSB award was similar, while the average VSP or ISP award with completed dual recoupment was smaller, approximately $27,000. Among cases in which recoupment has been completed, we observe that an average of $160,000 has been recouped from cases with DoD recoupment only and $180,000 has been recouped from retired pay from cases with dual recoupment. We estimate that VA recouped an additional $39,000 from VSI awards with dual recoupment.

As shown in the chart on the right (cases with ongoing recoupment), many of the values are similar: The average SSB award for cases with recoupment only from retired pay was again $43,000, and the average VSP or ISP award with recoupment only from retired pay was $42,000. Approximately half of the amount of recoupment is outstanding (i.e., has not yet been recouped) among cases with recoupment only from retired pay.

Among ongoing cases with dual recoupment, the SSB award size was again similar to the award sizes for the other groups, but the average VSP or ISP award was slightly higher, at $47,000. Among cases with dual recoupment, VA recoupment accounts for two-thirds to three-fourths of all recoupment that has occurred to date. An average of approximately $70,000–$80,000 of VSI awards was recouped from retired pay from cases with ongoing recoupment, although this likely reflects only a fraction of the total amount recouped to date. Among cases with ongoing VSI recoupment, we estimate that approximately $58,000 is outstanding to be recouped from retired pay. VA recoupment accounts for an additional $28,000, which is only a small fraction (26–28 percent) of total recoupment observed between 2013 and 2020. The reason VA recoupment of VSI represents a smaller share of total recoupment is that DFAS offsets only the amount of VADC received while VSI annuity payments are ongoing, while DFAS is required to recoup the full gross amount of VSI from retired pay (less any amount already recouped from VA). Because VSI awards can be much larger than SSB, VSP, or ISP, recoupment usually takes longer than eight years, which is the span of years we have available for this analysis.[10] As a result, we only observe a snapshot of ongoing recoupment for VSI.

In Table 4.3, we aggregate the individual amounts recouped from retired pay and VADC and the estimated amount outstanding across all veterans experiencing recoupment from retired pay. Between 2013 and 2020, approximately $116 million of VSP or ISP, $184 million of SSB, and $247 million of VSI was recouped from retired pay. Among veterans who experienced recoupment from both VADC and retired pay, an additional $42 million of VSP or ISP, $46 million of SSB, and an estimated $36 million of VSI was recouped from VADC. These numbers reflect the amount recouped from VA among veterans who fall in the intersection of the two circles in Figure 1.1 and are a fraction of the totals reported in Table 4.3. Because we have incomplete data on recoupment to date for VSI, the $36 million estimate is a lower-bound estimate; the true amount recouped from VA is likely higher. Finally, we estimate that approximately $18 million of VSP or ISP is subject to recoupment from retired pay but has not yet been recouped. The estimated outstanding recoupment amounts for SSB and VSI are $88 million and $135 million, respectively.

TABLE 4.3

Aggregated Total Amount Recouped from Military Retired Pay, 2013–2020

Separation Pay Type	Total Recouped from Retired Pay	Total Recouped from VA	Total Outstanding Recoupment from Retired Pay
VSP or ISP	$115.8 million	$41.5 million	$18.8 million
SSB	$183.5 million	$46.2 million	$87.6 million
VSI	$247.2 million	$35.6 million[a]	$135.0 million

SOURCE: Data from 2013–2020 provided to the authors by DFAS.
[a] Estimates of total remitted to VA for VSI are based on cases in which the VSI annuity is complete.

---

[10] See the appendix for more details on the duration of recoupment of separation pay from retired pay.

CHAPTER FIVE

Findings and Discussion

We conclude our report with a summary of the findings and a discussion of the implications of the analysis for policies related to recoupment.

Findings

Table 5.1 shows a summary of our results with respect to the two questions posed by Congress. In considering these results, it is important to recall several important data limitations. In particular, the data did not permit us to compute the total number of veterans who experienced and completed VA recoupment of either SSB or separation pay before 2013, because VA

TABLE 5.1

Summary of Results for the Two Congressional Questions

	Question 1	Question 2		
Separation Pay Type	Count of Veterans	Average Recouped per Person	Total Recouped	Estimated Total Recouped in 2021
VSP or ISP (2013–2020)	72,206	$19,700	$1.4 billion	$67 million
SSB (2013–2020)	2,651	$25,700	$68 million	$730,000
VSI (1992–2020)	4,701	$53,000	$177 million	$9.6 million

SOURCE: Data from 1992 to 2020 provided to the authors by DFAS; data from 2013–2020 provided to the authors by VA.
NOTE: VSI computations for Question 2 are based on cases in which recoupment has been completed. Estimates of the total amount to be recouped in 2021 are based on cases for which we observed recoupment at the time that the data were collected. The total amount recouped for VSI is based on cases in which VSI payments are complete.

data systems underwent a major change at that time. Furthermore, we were unable to compute either the share of service members who received SSB or VSP/ISP who ever experienced VA recoupment or the share of veterans receiving VADC who ever experienced VA recoupment. Finally, because of differences in the structure of available data, we were unable to determine the total amount recouped before December 2020 for individuals still experiencing recoupment of VADC from VSI.

With respect to the first question, we found that the number of veterans who experienced recoupment because of the receipt of separation pay is much larger than the number with recoupment of either VSI or SSB. Separation pay programs, specifically ISP and VSP, are ongoing, while both the VSI and SSB programs ended in 2001. Although it is possible that some veterans who received these payments in the past might begin receiving VADC in the future and thus become subject to recoupment in the future, we suspect that to be unlikely. With respect to the second question, we found that, on average, veterans have had $19,700 or $25,700 withheld because of recoupment of separation pay or SSB, respectively, but have had $53,000 withheld because of recoupment of VSI. In aggregate, over the eight-year period from 2013 to 2020, a total of $1.4 billion of VADC payments were withheld because of the recoupment of separation pay, a far larger figure than the $68 million withheld because of the recoupment of SSB. DFAS has withheld a minimum of $177 million of VSI in aggregate because of the receipt of VADC.[1]

Discussion

In the past, members of Congress have argued that VADC and DoD separation pay are not duplicative and that recoupment is a "bureaucratic mistake" (Clark, 2017). Some have argued for the elimination of recoupment of separation pay. But elimination of this policy raises several issues.

---

[1] Recall that this is a lower-bound estimate, because we were unable to estimate the amount recouped to date from cases in which the VSI annuity is still being paid.

Cost to the Government

Our analysis summarized in Table 5.1 indicates that, had recoupment of VSP or ISP not occurred, $1.4 billion more of VADC would have been paid between 2013 and 2020, or $67 million per year. The elimination of recoupment of SSB would have separately added $68 million total, or approximately $730,000 per year, to these figures. The table also indicates that had recoupment from VSI due to the receipt of VADC not occurred, $177 million of VSI payments would have been made since 1992 among cases that were completed as of December 2020. We estimate that an additional $9.6 million of VSI will be recouped in 2021.[2]

Another consideration related to the cost to the government is that recoupment is based on the nominal value of the separation pay at the time the award was made, without interest. This results in a lower amount recouped in real terms, particularly for VSI and SSB for which the award values were determined and initial payments were made at least 20 years ago. In this respect, recoupment differs from other debt collection, in which interest typically accrues over time (DoD Financial Management Regulation, Vol. 7B, Chapter 2; see DoD, 2021a).

Other Types of Separation Pay Recoupment

The cost estimates in Table 5.1 reflect the amount of recoupment of separation pay because of VADC, as requested by Congress. However, recoupment of separation pay can occur for other reasons, and other types of separation pay may be subject to recoupment. Eliminating recoupment of separation pay because of receipt of VADC without eliminating recoupment under these other circumstances could raise concerns about the equitable treatment of veterans. For example, elimination of recoupment because of receipt of VADC without also eliminating recoupment because of receipt of military retired pay might be viewed as inequitable given that some veterans

---

[2] Note that our estimates of cost to the government are not estimates of cost should Congress repeal recoupment of separation pay. Much would depend on the details of such a repeal. For example, the cost would depend on whether the repeal was retroactive, so the government would be required to refund past recoupment, and whether such a refund would apply to survivors of deceased veterans. Analysis of alternative policies with respect to repealing recoupment was beyond the scope of this project.

can experience both types of recoupment, as shown by the intersection of the two circles in Figure 1.1.

In Table 5.2, we summarize our findings from the exploratory analysis of recoupment from retired pay discussed in Chapter Four. These estimates suggest that eliminating recoupment of separation pay from military retired pay would increase costs. The aggregated total of recoupment of all types of separation pay from retired pay is still smaller than the aggregated total of all recoupment from VADC, but the amount of outstanding recoupment from VSI and SSB is larger and could increase in the coming years as more veterans reach retirement age.

Table 5.2 provides important perspective on the differences in recoupment from VADC summarized in Table 5.1 and from military retired pay. First, the number of veterans subject to recoupment differs greatly. While the number of veterans with recoupment of VSP or ISP from retired pay is much lower than the numbers for recoupment from VADC, the number of veterans with recoupment of SSB from retired pay is higher. Because more veterans who took SSB may reach qualification for retirement in the coming years, the total number of veterans subject to recoupment could increase even further (although not indefinitely, given that SSB ended in 2001). A similar increase could occur in cases of recoupment of VSI from military retired pay, although the increase has not been as dramatic to date, perhaps because fewer members took VSI than took SSB.

TABLE 5.2

Summary of Results for Recoupment from Military Retired Pay, 2013–2020

Separation Pay Type	Question 1	Question 2	
	Count of Veterans	Average Recouped per Person	Total Recouped from Retired Pay or VADC
VSP or ISP	5,403	$14,165	$157.3 million
SSB	7,154	$15,989	$229.7 million
VSI	2,706	$73,757	$282.8 million

SOURCE: Data from 2013–2020 provided to the authors by DFAS.
NOTES: Estimates of total remitted to VA for VSI are based on cases in which the VSI annuity is complete. The final column represents the sum of the two left columns in Table 5.2.

Second, differences in the numbers and, in the case of VSI, the amounts subject to recoupment lead to differences in the aggregated amount that veterans would receive but for recoupment. Although the average per-person value of VSP or ISP is higher among veterans experiencing recoupment from military retired pay, the aggregated total is much lower because fewer veterans with VSP or ISP experience recoupment from retired pay. However, the totals for SSB are higher, both because average SSB amounts are higher and because there are more veterans with SSB who experience recoupment from retired pay. The average per-person VSI recoupment from retired pay is much higher given that DFAS recoups the full gross value of VSI while DFAS only reduces the amount of VSI by any VADC received. Even though the numbers of veterans experiencing VSI recoupment from retired pay are slightly smaller than on the VA side, the aggregated total is larger because of the larger amount subject to recoupment.

Disability Severance Pay

While the focus of our study has been on VSI, SSB, and separation pay (VSP and ISP)—the categories of separation pay identified by Congress—another separation benefit for which recoupment occurs is DoD DSP. Service members with fewer than 20 YOS who are found unfit for service and receive a DoD disability rating of less than 30 percent may receive a lump-sum severance payment from DoD. This severance payment may be subject to recoupment by VA if the veteran also receives VADC. Eliminating recoupment of separation pay due to the receipt of VADC without also eliminating recoupment of DoD DSP due to the receipt of VADC could raise concerns of unequal treatment of veterans subject to recoupment.

Estimating the number of veterans receiving DSP who experience recoupment and the amount of money they would have received in the absence of recoupment was beyond the scope of our study. That said, Rennane et al., forthcoming, shows that about 8,000, or about 25 percent, of service members who were found unfit for service and received a DoD disability rating received DSP in 2015. The majority of these 8,000 service members will also receive a VA disability rating, although not all experience recoupment of benefits if their unfitting conditions are found to be combat-related.[3] These estimates suggest that a sizable number of veterans might be potentially affected by elimination of recoupment due to the receipt of both VADC and DoD DSP with a sizable associated cost.

Duplication of Benefits

Those who support eliminating separation pay recoupment, especially recoupment of separation pay due to the receipt of VADC, argue that separation pay and VADC have distinct purposes that are not duplicative. Consequently, they argue that the logic of double-dipping does not apply. Separation pay is a tool to facilitate the voluntary or involuntary separation of military personnel when it is no longer for the good of the service for the member to continue to serve. Often, the circumstances surrounding the offer of separation pay are beyond a service member's control, such as the drawdown of forces by the services or the result of a service-connected injury that makes a member unfit for continued service. As shown in Chapter Two, the formula for computing separation pay benefits sets the level of benefits based on career metrics at the time of separation, such as years of military service and pay grade. Thus, the amount of separation pay reflects the amount of service performed by the member.

Military retired pay, like separation pay, is also based on career metrics; the military retired pay formula depends on YOS and basic pay, and basic pay, in turn, depends on pay grade and YOS. In addition, like separation pay, military retired pay helps serve as a transition benefit as members transition from military service to the civilian labor market.

In contrast to either separation pay or military retired pay, VADC is not based on career metrics, such as YOS and rank. Instead, VADC depends on the extent of the service-connected disability, or VA disability rating, and family structure, such as number of dependents. VADC is intended to replace the decrease in civilian earnings resulting from the service member's disability, on average.

---

[3] We note that not all veterans who receive DoD severance pay and are eligible for VADC will experience recoupment. For example, if the disability was incurred in a combat zone, the member might be eligible for Combat Related Special Compensation that offsets recoupment (DoD, 2020).

The laws and regulations surrounding the prohibition of double-dipping are usually explained in terms of a prohibition from receiving payments from federal agencies for the same purpose.[4] It is clear from intent and the formulas for setting benefits that separation and retired pay are arguably duplicative in some ways, while VADC is distinct from either separation pay or retired pay.

However, the regulation governing this matter does not mention either duplication of intent or duplication of formulas. In particular, 38 C.F.R. 3.700 states,

> Not more than one award of pension, compensation, or emergency officers', regular or reserve retirement pay will be made concurrently to any person based on his or her own service except as provided in §3.803 relating to naval pension and §3.750(c) relating to waiver of retirement pay.

Thus, the key issue is whether the two sources of compensation are based on the same period of service.

To eliminate recoupment would require either a change in the wording of 38 U.S.C. 5304(a) or the creation of a new benefit that would offset recoupment of separation pay for qualified veterans. For example, such a benefit might work like Combat Related Special Compensation or Concurrent Retirement and Disability Payments that provide qualified veterans who receive DoD disability compensation a payment that offsets the reduction in their DoD compensation due to the receipt of VADC. Assessment of such a new benefit, including the administrative costs of the benefit, was beyond the scope of our study.

Wrap-Up

In sum, we find that nearly 75,000 veterans have experienced VA recoupment of SSB, VSP, or ISP since 2013, and just over 4,700 veterans have ever experienced recoupment of VSI due to receipt of VADC. The vast majority—

---

[4] See, for example, Kamarck, 2019.

more than 72,000—of these veterans received separation pay. On average, veterans with SSB, VSP, or ISP would have received between $20,000 and $36,000 but for recoupment of this pay, and veterans with VSI would have just over $50,000 in additional compensation but for recoupment. Because the vast majority of cases with VA recoupment are VSP or ISP, these cases represent the largest amount of money in aggregate, with nearly $1.4 billion dollars recouped between 2013 and 2020. We estimate that the aggregated amount of recoupment for SSB is $68 million between 2013 and 2020. Furthermore, we estimate that the aggregated amount of VSI recoupment among completed cases is $177 million.

These analyses provide answers to the questions posed by Congress but also raise several other questions that could be explored in future research. Although we present preliminary analyses of recoupment of military retired pay, additional work needs to be done to analyze the potential implications of recoupment of other types of separation pay, including disability severance. Furthermore, with additional data, future analyses could provide a better understanding of the characteristics of service members and veterans who are affected by recoupment, including their age, gender, location, and characteristics of their time in service. And finally, additional data could enable more-robust projections of how the number of service members experiencing recoupment, and the amount of money that members would receive but for recoupment, could change in the future.

Appendix

This appendix includes additional analyses related to the recoupment of separation pay from military retired pay. These analyses support the discussion of recoupment of military retired pay in Chapters Four and Five.

Figure A.1 shows the number of veterans who experienced recoupment of VSP or ISP, SSB, or VSI due to receipt of military retired pay in each year between 2013 and 2020. While the number of veterans with recoupment of VSP or ISP from military retired pay declines over this period, the number of veterans experiencing recoupment of SSB and VSI from retired pay increases. This increase is likely because many veterans who took these special separation pays between 1992 and 2001 are now reaching retirement age.

FIGURE A.1

Number of Veterans with Any Recoupment of Separation Pay (VSP or ISP), SSB, or VSI Due to Receipt of Military Retired Pay over Time, 2013–2020

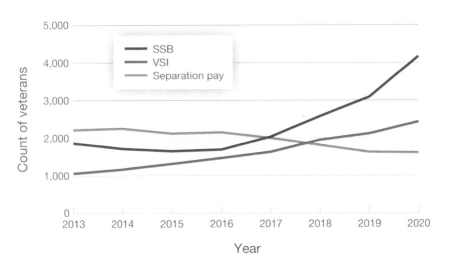

SOURCE: Data from 2013–2020 provided to the authors by DFAS.
NOTE: The label "Separation pay" refers to ISP or VSP.

Next, we conduct some analyses of the duration of recoupment among cases for which we observe the full period of recoupment in our data extract. In other words, we restrict the sample in these analyses to individuals who began recoupment during or after 2013 and completed recoupment by the end of 2020. This corresponds to the blue bars shown in Figure A.2. We observe the full period of recoupment for approximately 38 percent of the cases in our file. Figure A.2 shows that the majority of these cases were SSB (2,335) and VSP/ISP (2,550), while we observe the full period of recoupment for only 796 cases of VSI.

Figure A.3 shows the average per-person amount recouped from each type of separation pay among completed cases for which we observe the full period of recoupment in our data extract. Among cases with recoupment only from retired pay, the average per-person amount recouped from VSP/ISP was just over $30,000, the average amount recouped from SSB was just over $40,000, and the average amount recouped from VSI was approximately $150,000. Among cases with recoupment from retired pay

FIGURE A.2

Number of Veterans with Full Recoupment from Military Retired Pay Observed Between 2013 and 2020, by Type of Separation Pay

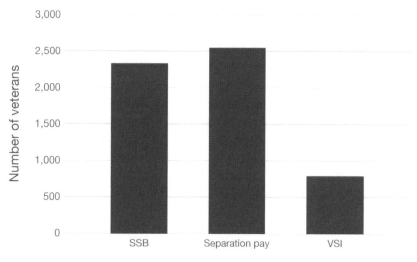

SOURCE: Data from 2013 to 2020 provided to the authors by DFAS.
NOTE: The label "Separation pay" refers to ISP or VSP.

FIGURE A.3

Average Per-Person Amount of Separation Pay, SSB, and VSI Recouped, Among Awards with Completed Recoupment from Military Retired Pay

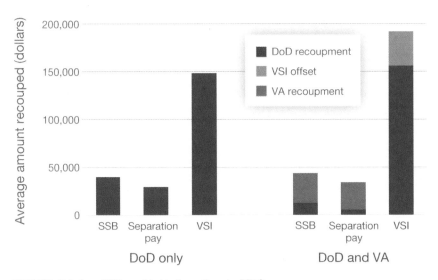

SOURCE: Data from 2020 provided to the authors by DFAS.
NOTE: Estimate of VSI offset from VA is based on recoupment of VSI among cases in which VSI annuity is complete, which account for approximately two-thirds of all cases with dual recoupment of VSI from DoD and VA.

and VADC, the average per-person amount recouped from VSP/ISP was $35,000, the average amount recouped from SSB was $44,000, and the average amount recouped from VSI was $191,000. For cases with dual recoupment, approximately 80 percent of recoupment of SSB and 70 percent of recoupment of VSP/ISP occurred from VA. By contrast, only 18 percent of recoupment of VSI was a deduction for VADC.

Notably, the total amounts subject to recoupment for SSB and ISP/VSP are similar (although slightly smaller, since smaller awards are recouped more quickly) to what we observe for ongoing cases in the data in Figure A.3, suggesting that the completed cases might be a reasonable proxy for what ongoing cases will look like. By contrast, the recoupment that occurred before 2013 and after 2020 is censored for VSI cases for which we do not observe full recoupment between 2013 and 2020, so we are unable to compare the completed cases with the ongoing cases. However, because ongoing awards are likely to be larger, the averages shown in Figure A.3 likely present a lower bound on total recoupment from VSI.

Using this sample for which we observe full recoupment, we conduct some analyses to determine how long recoupment typically takes. Figures A.4–A.6 provide tabulations of the number of years recoupment took for cases in which SSB, VSP/ISP, and VSI were fully recouped. On average, it took three years to fully recoup SSB and VSP/ISP, although the figures show that there is still a range, with some awards being recouped within one year and others taking eight years to be fully recouped. The distribution looks quite different for VSI, which took an average of six years to fully recoup. Figure A.7 shows the cumulative distribution of the number of years required to complete recoupment within this subsample for which we observe full recoupment. Again, the difference in the distribution for VSI is striking, suggesting that VSI awards take a much longer time to be fully recouped.

FIGURE A.4

Count of Cases of Recoupment of SSB from Military Retired Pay, Based on the Number of Years Required to Fully Recoup SSB

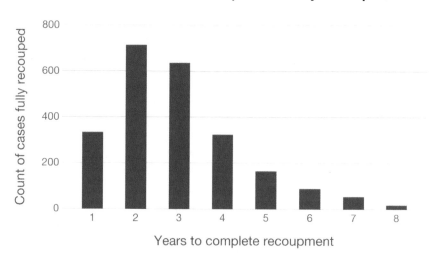

Years to complete recoupment

SOURCE: Data from 2013 to 2020 provided to the authors by DFAS.
NOTE: The sample for this analysis is restricted to cases for which both the beginning and end of recoupment occur between 2013 and 2020 (N = 5,681).

FIGURE A.5

Count of Cases of Recoupment of ISP or VSP from Military Retired Pay, Based on the Number of Years Required to Fully Recoup Separation Pay

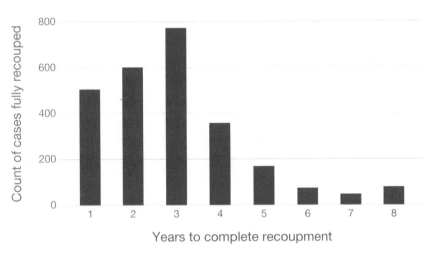

SOURCE: Data from 2013 to 2020 provided to the authors by DFAS.
NOTE: The sample for this analysis is restricted to cases for which both the beginning and end of recoupment occur between 2013 and 2020 (N = 5,681).

Because the total awards for completed cases look similar to the total awards for ongoing cases with recoupment of SSB and VSP/ISP, these analyses shed some light on how long recoupment likely takes even for the ongoing cases. The VSI sample, by contrast, is unique because it is unusual for VSI recoupment to be completed within eight years.

FIGURE A.6

Count of Cases of Recoupment of VSI from Military Retired Pay, Based on the Number of Years Required to Fully Recoup VSI

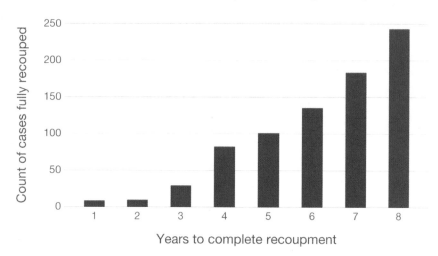

Years to complete recoupment

SOURCE: Data from 2013 to 2020 provided to the authors by DFAS.
NOTE: The sample for this analysis is restricted to cases for which both the beginning and end of recoupment occur between 2013 and 2020 (N = 5,681).

FIGURE A.7

Cumulative Distribution of the Number of Years Required to Recoup Separation Pay from Military Retired Pay, by Pay Type

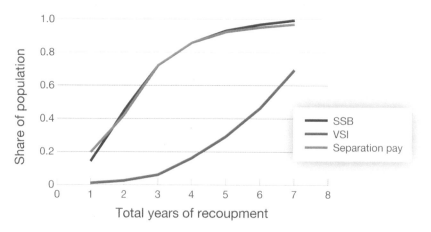

Total years of recoupment

SOURCE: Data from 2013 to 2020 provided to the authors by DFAS.
NOTE: The sample for this analysis is restricted to cases for which both the beginning and end of recoupment occur between 2013 and 2020 (N = 5,681).

Abbreviations

DFAS	Defense Finance and Accounting Service
DoD	U.S. Department of Defense
DoDI	Department of Defense Instruction
DSP	disability severance pay
FY	fiscal year
ISP	involuntary separation pay
NDAA	National Defense Authorization Act
RISP	reservists involuntary separation pay
SSB	special separation benefits
VA	U.S. Department of Veterans Affairs
VADC	VA Disability Compensation
VSI	voluntary separation incentive
VSP	voluntary separation pay
YOS	years of service

References

Asch, Beth J., James Hosek, and Michael G. Mattock, *Toward Meaningful Military Compensation Reform: Research in Support of DoD's Review*, Santa Monica, Calif.: RAND Corporation, RR-501-OSD, 2014. As of November 16, 2020:
https://www.rand.org/pubs/research_reports/RR501.html

Clark, James, "A Law Against 'Double Dipping' Is Forcing Disabled Vets to Pay Back Separation Benefits," *Task & Purpose*, May 4, 2017. As of July 15, 2021:
https://taskandpurpose.com/veterans-benefits/
veterans-disability-compensation-separation-pay-duplication-of-benefits/

Code of Federal Regulations, Title 38, Pensions, Bonuses, and Veterans' Relief; Chapter I, Department of Veterans Affairs; Part 3, Adjudication; Subpart A, Pension, Compensation, and Dependency and Indemnity Compensation; Subjgrp, Concurrent Benefits and Elections; Section 3.700, General. As of July 15, 2021:
https://www.govinfo.gov/app/details/CFR-2010-title38-vol1/
CFR-2010-title38-vol1-sec3-700

Defense Finance and Accounting Service, "Disability Severance Pay," webpage, last updated July 29, 2016. As of November 16, 2020:
https://www.dfas.mil/retiredmilitary/plan/separation-payments/
disability-severance-pay/

Department of Defense Instruction 1332.29, *Involuntary Separation Pay (Non-Disability)*, Washington, D.C.: U.S. Department of Defense, March 3, 2017. As of November 16, 2020:
https://www.esd.whs.mil/Portals/54/Documents/DD/issuances/dodi/
133229p.pdf

Department of Defense Instruction 1332.43, *Voluntary Separation Pay (VSP) Program for Service Members*, Washington, D.C.: U.S. Department of Defense, November 28, 2017. As of November 16, 2020:
https://www.esd.whs.mil/Portals/54/Documents/DD/issuances/dodi/
133243p.pdf?ver=2017-11-28-102301-607

DFAS—*See* Defense Finance and Accounting Service.

DoD—*See* U.S. Department of Defense.

DoDI—*See* Department of Defense Instruction.

Kamarck, Kristy N., *Concurrent Receipt: Background and Issues for Congress*, Washington, D.C.: Congressional Research Service, R40589, updated January 17, 2019. As of May 5, 2021:
https://fas.org/sgp/crs/misc/R40589.pdf

Office of the Under Secretary of Defense (Comptroller), "DoD Budget Request," webpage, undated. As of July 21, 2021: https://comptroller.defense.gov/Budget-Materials/

Public Law 102-190, National Defense Authorization Act for Fiscal Years 1992 and 1993, December 5, 1991. As of July 15, 2021: https://www.govinfo.gov/content/pkg/STATUTE-105/pdf/ STATUTE-105-Pg1290.pdf

Public Law 102-484, National Defense Authorization Act for Fiscal Year 1993, October 23, 1992. As of July 15, 2021: https://www.congress.gov/ search?q=%7B%22search%22%3A%22cite%3APL102-484%22%7D

Public Law 116-92, National Defense Authorization Act for Fiscal Year 2020, December 20, 2019. As of August 27, 2021: https://www.govinfo.gov/content/pkg/PLAW-116publ92/pdf/ PLAW-116publ92.pdf

Rennane, Stephanie, Beth J. Asch, Michael G. Mattock, Heather Krull, Douglas C. Ligor, Michael Dworsky, and Jonas Kempf, *U.S. Department of Defense Disability Compensation Under a Fitness-for-Duty Evaluation Approach*, Santa Monica, Calif.: RAND Corporation, RR-A1154-1, forthcoming.

Tilghman, Andrew, "Vets Dismissed by VA Payback Rules on Separation Pay," *Military Times*, October 18, 2015. As of July 15, 2021: https://www.militarytimes.com/veterans/2015/10/18/ vets-dismayed-by-va-payback-rules-on-separation-pay/

U.S.C.—*See* U.S. Code.

U.S. Code, Title 10, Armed Forces; Subtitle A, General Military Law; Part II, Personnel; Chapter 59, Separation; Section 1174, Separation Pay upon Involuntary Discharge or Release from Active Duty. As of July 15, 2021: https://www.govinfo.gov/app/details/USCODE-2011-title10/ USCODE-2011-title10-subtitleA-partII-chap59-sec1174

U.S. Code, Title 10, Armed Forces; Subtitle A, General Military Law; Part II, Personnel; Chapter 59, Separation; Section 1174a, Special Separation Benefits Programs. As of July 15, 2021: https://www.govinfo.gov/app/details/USCODE-2011-title10/ USCODE-2011-title10-subtitleA-partII-chap59-sec1174a

U.S. Code, Title 10, Armed Forces; Subtitle A, General Military Law; Part II, Personnel; Chapter 59, Separation; Section 1174h, Coordination with Retired or Retainer Pay and Disability Compensation. As of July 15, 2021: https://www.govinfo.gov/content/pkg/USCODE-2011-title10/pdf/ USCODE-2011-title10-subtitleA-partII-chap59-sec1174.pdf

U.S. Code, Title 10, Armed Forces; Subtitle A, General Military Law; Part II, Personnel; Chapter 59, Separation; Section 1175a, Voluntary Separation Pay and Benefits. As of July 15, 2021:
https://www.govinfo.gov/app/details/USCODE-2011-title10/
USCODE-2011-title10-subtitleA-partII-chap59-sec1175a

U.S. Code, Title 10, Armed Forces; Subtitle A, General Military Law; Part II, Personnel; Chapter 61, Retirement or Separation for Physical Disability; Section 1212, Disability Severance Pay. As of July 15, 2021:
https://www.govinfo.gov/app/details/USCODE-2011-title10/
USCODE-2011-title10-subtitleA-partII-chap61-sec1212

U.S. Code, Title 38, Veterans' Benefits; Part IV, General Administrative Provisions; Chapter 53, Special Provisions Relating to Benefits; Section 5304, Prohibition Against Duplication of Benefits. As of July 15, 2021:
https://www.govinfo.gov/app/details/USCODE-2011-title38/
USCODE-2011-title38-partIV-chap53-sec5304

U.S. Department of Defense, *Military Compensation Background Papers: Compensation Elements and Related Manpower Cost Items: Their Purposes and Legislative Backgrounds*, 8th ed., Washington, D.C., July 2018. As of November 16, 2020:
https://www.loc.gov/rr/frd/pdf-files/Military_Comp-2018.pdf

———, "Recoupment of Separation Pay," Financial Management Regulation, DoD 7000.14-R, Vol. 7B, Chapter 4, March 2020.

———, "Collection of Debts," Financial Management Regulation, DoD 7000.14-R, Vol. 7B, Chapter 2, January 2021a. As of May 11, 2021:
https://comptroller.defense.gov/Portals/45/documents/fmr/
archive/07barch/07b0602.pdf

———, "Recoupment of Separation Pay," Financial Management Regulation, DoD 7000.14-R, Vol. 7B, Chapter 4, January 2021b. As of March 24, 2021:
https://comptroller.defense.gov/Portals/45/documents/fmr/
current/07b/07b_04.pdf

U.S. Department of Veterans Affairs, "M21-1, Part III, Subpart v, Chapter 4, Section B - Recoupment of Separation Benefits," webpage, undated. As of November 16, 2020:
https://www.knowva.ebenefits.va.gov/system/templates/selfservice/
va_ssnew/help/customer/locale/en-US/portal/554400000001018/
content/554400000014245/M21-1,-Part-III,-Subpart-v
,-Chapter-4,-Section-B---Recoupment-of-Separation-Benefits

VA—*See* U.S. Department of Veterans Affairs.

Warner, John T., and Saul Pleeter, "The Personal Discount Rate: Evidence from Military Downsizing Programs," *American Economic Review*, Vol. 91, No. 1, March 2001, pp. 33–53.